D1527725

CRC Series in
Materials Science and Technology

Series Editor
Brian Ralph

Control of Microstructures and Properties in Steel Arc Welds
Lars-Erik Svensson

The Extraction and Refining of Metals
Colin Bodsworth

The Quantitative Description of the Microstructure of Materials
K.J. Kurzydłowski and Brian Ralph

Grain Growth and Control of Microstructure and Texture in Polycrystalline Materials
Vladimir Novikov

Corrosion Science and Technology
D. E. J. Talbot and J. D. R. Talbot

Image Analysis: Applications in Materials Engineering
Leszek Wojnar

image analysis

Applications in Materials Engineering

Leszek Wojnar

CRC Press
Boca Raton London New York Washington, D.C.

Library of Congress Cataloging-in-Publication Data

Wojnar, Leszek.
Image analysis applications in materials engineering / Leszek Wojnar.
p. cm. -- (Materials science and technology series)
Includes bibliographical references and index.
ISBN 0-8493-8226-2 (alk. paper)
1. Materials--Testing. 2. Image analysis. 3. Image processing--Digital techniques. I. Title. II. Series: Materials science and technology (Boca Raton, Fla.)
TA410.W65 1998
621.1'1—dc21 98-34435
CIP

Direct all inquiries to CRC Press LLC, 2000 Corporate Blvd., N.W., Boca Raton, Florida 33431.

International Standard Book Number 0-8493-8226-2
Library of Congress Card Number 98-34435
Printed in the United States of America 1 2 3 4 5 6 7 8 9 0
Printed on acid-free paper

Preface

Many of my friends complained that all the books on image analysis were prepared for mathematicians rather than for laboratory workers. It was an impulse to issue in 1994, in collaboration with Mirosław Majorek, a simple textbook on computer-aided image analysis, obviously written in Polish. Some of my foreign colleagues looked at this book, appreciated its graphic form and... suggested, that it should be published in English.

My first reaction was that it was not worthy enough. Surprisingly, they answered with a very tempting argument: you started to write this book because you did not find appropriate ones on the market. Maybe yes... So, now you have my English version in your hands and I would like to point out its three main properties, that can be important for you, as a reader:

- it is devoted really to applications. So, you will not find a systematic description of image processing operations. Instead, you can look for a certain problem - for example, grain boundary detection - and get immediately, possibly a full solution to this problem
- it is written in a very simple manner and illustrated with numerous pictures that will help you to understand it. Probably, many items can be understood after studying only the illustrations. But do not worry about the text - I avoid equations whether possible
- all the examples were processed by myself and thoroughly explained. You will not find incomplete explanations, cited from other works. It may happen that my solution is not the optimum one, but it always works. You will know how to repeat it on your own equipment and I hope, my book will inspire you to experiment with your apparatus.

Probably nobody is able to face a challenge such as writing a technical book without significant help from others. I am not an exception, either. So, I would like to express my sincere thanks to all those, who helped me, even if I am unable to cite their names - simply, the list would be too long. But among these generous persons there are a few I must list here.

First, I am really indebted to Brian Ralph, who generously agreed to undertake the burden of improving my English. Second, I would like to thank Christopher Kurzydłowski, who encouraged me to start the whole project, for his support. I have to point out that my understanding of image analysis would be much, much less without the support from my French colleagues, Jean Serra and, especially, Jean-Louis Chermant. I would like to acknowledge also all my friends who gave me their images which were used to illustrate the algorithms. And last but not least - many thanks to Bruno Lay and Gervais Gauthier from ADCIS/AAI, who delivered free of charge, the newest version of their image analysis software to use for processing all the examples.

Leszek Wojnar

Cracow, August 1998

The Author

Leszek Wojnar, D.Sc., is an Associate Professor at Cracow University of Technology, Poland. His research interests are in image analysis, stereology and materials engineering. He is affiliated with the Polish Society for Materials Engineering, the International Society for Stereology and the Polish Society for Stereology.

His Ph.D. thesis obtained at the Institute of Materials Science, Cracow University of Technology (1985) was dedicated to the role of nodular graphite in the fracture process of cast iron. Dr Wojnar is known for his innovative methods in teaching oriented towards problem solving. In recent years (1990-1997) he has worked on various aspects of the application of computer technology in materials science, including image analysis and development of the software for weldability assessment. Dr Wojnar has participated as a member to various advisory boards of the congresses in stereology (Freiburg, 1989, Irvine, California, 1991, Prague, 1993 and Warsaw, 1997). He was an invited lecturer in Freiburg and Warsaw and worked as editor for many conference proceedings.

Dr Wojnar has published more than 50 articles in periodicals and conference proceedings and has published two books (in Polish): *Computerized Image Analysis* with M. Majorek (1994) and *Application of Computational Methods in Weldability Assesment* with J. Mikula (1995). His work *Principles of Quantitative Fractography* (1990) issued at the Cracow University of Technology was the first such complete monograph in Poland and gave him the D.Sc. position.

An unusual accomplishment and one which stems from the existing financial climate surrounding Cracow University is that he developed his own private laboratory on image analysis which now works in conjunction with universities throughout Poland. Recent works of this small laboratory are devoted mainly to applications in medicine.

Acknowledgments

I would like to express my sincere thanks to all my friends and colleagues who nicely allowed me to use their images as illustration material in this book. These generous helpers were (in alphabetical order):

- Jacek Chrapoński (Figures 4.11, 4.12a and 4.13a)
- Aleksandra Czyrska-Filemonowicz (Figuress 4.29, 4.31a, 5.6 and 5.7)
- Wiesław Dziadur (Figures 2.3 and 4.23a, b)
- Marek Faryna (Figures 4.4a and 4.23d)
- Gervais Gauthier (Figure 4.16a)
- Jan Głownia (Figure 4.27)
- Gabriela Górny (Figures 2.18a, 2.28a and 4.28a)
- Krzysztof Huebner (Figures 2.32, 3.1, 3.2, 3.3, 3.4, 3.5 and 3.6)
- Anna Kadłuczka (Figure 4.3a)
- Jan Kazior (Figures 2.10, 2.11a and 7.2)
- Krzysztof Kurzydłowski (Figure 4.24)
- Anita Myalska-Olszówka (Figure 5.9)
- Carl Redon (Figure 4.40a)
- Kazimierz Satora (Figures 4.23e, 5.14a and 5.15a)
- Janusz Szala (Figures 3.7, 3.8, 3.9, 4.17a and 4.18b)
- Adam Tabor (Figures 2.9, 3.10 and 4.26)
- Roman Wielgosz (Figures 2.13a, 2.15a and 4.30a).

Thank you.

Contents

Chapter one

Introduction

1.1 Digital images and their analysis versus the human vision system

During the last ten years we observed a tremendous expansion of more and more powerful personal computers and development of user-friendly, graphically oriented operating systems. The computing power of commercial machines doubles in approximately one to two years, and even more powerful computers are used on a laboratory scale. Obviously, the most advanced computer is worth nothing without appropriate software. The unprecedented market success of general purpose software developers forced numerous smaller companies to look for niche applications, very often connected with computer graphics. The availability of frame grabbers, together with the wide range of video cameras, allows the computer to see images and induces the temptation to try simulation of the human vision system. As a consequence, a good deal of image analysis software is currently at hand. It allows many research workers to practice with tools previously available only for a limited group of specialists.

However, new tools also provide new problems, often caused by some misunderstanding. High resolution graphics allows one to produce photo-realistic effects, leading to impressive virtual reality products. One can walk through non-existent buildings, observe crash tests of newly designed but still non-existent cars, train surgeons on virtual patients, etc. Similar effects can be obtained on small scale in almost all personal computers and the appropriate software is commercially marketed.

Computerized graphical presentations, often demonstrated in real-time mode, are extremely impressive, especially for novices. They are used very frequently, particularly for advertising purposes. Unfortunately, such breathtaking spectacles in virtual reality may lead to a false impression that computers can do almost everything. Moreover, many people are disappointed and frustrated when trying to do anything on their own. Such a case is very common in image analysis applications, which work perfectly, but only on the test images. Let us try to find the reason for this situation.

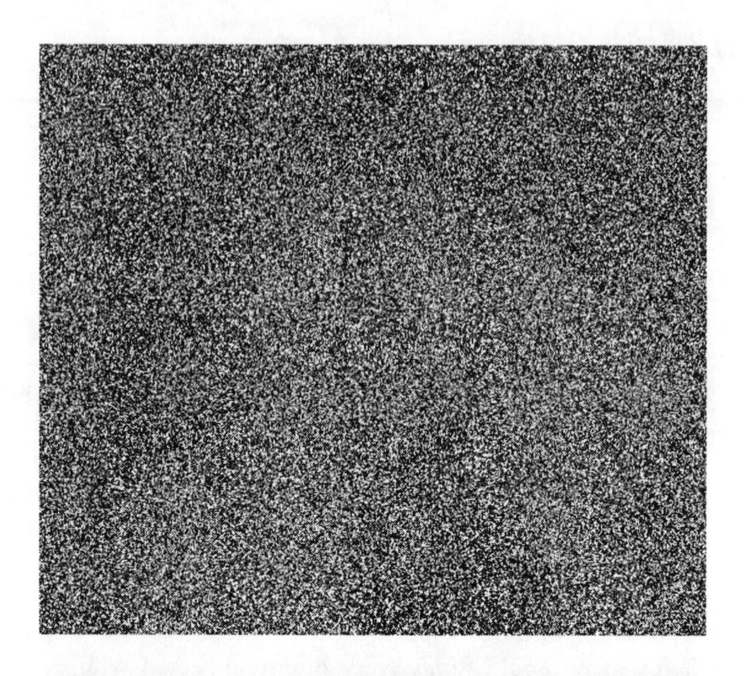

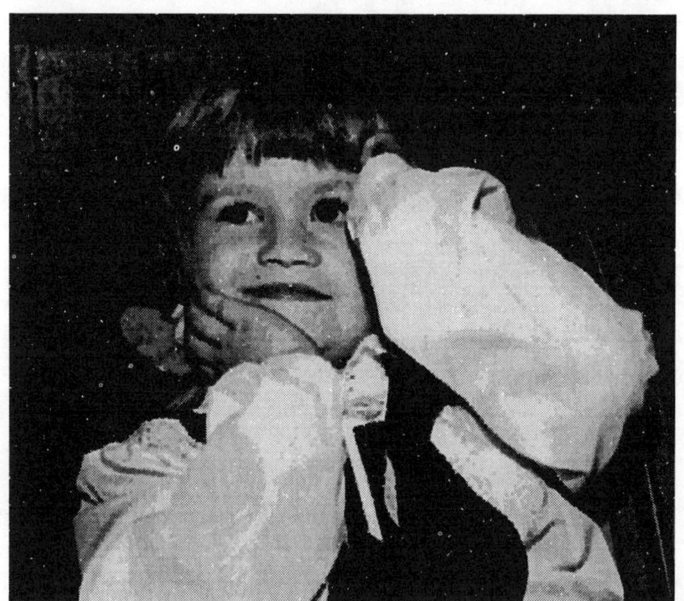

Fig. 1.1. The noisy image (up) contains some information which is totally invisible for human eyes but can be easily developed after proper application of very simple transformations (down). See text for more details.

Let us analyze the upper image in Fig. 1.1. It looks as uniformly noisy, perhaps with some brighter areas in the middle and lower right-hand corner. However, it is enough to apply two simple steps:

- minimum filter that converts any point in the image into its darkest neighbor (see Section 2.3 for more details)
- simple linear LUT transformation (see Section 2.2 for more details) in order to get optimum brightness and contrast.

After such a treatment we get the lower image shown in Fig. 1.1. Thanks to the simplicity of the transformations applied the whole process is very fast and impressive. We can add that for nearly every image it is possible to find an appropriate procedure that can extract features necessary for further analysis. The only problem is that usually we do not know

HOW TO DO IT?

In demonstration files this problem is already solved. Detailed analysis of the demo can give us some guidelines for our own cases, but it is never universal knowledge. So, the next and possibly the most important problem is

HOW MUCH TIME WILL WE SPEND TO FIND THE PROPER SOLUTION?

It depends on our experience, type of image to be analyzed, etc. However, in general, to find the acceptable solution takes much more time than we expect.

It seems a paradox that we have an extremely powerful image analysis program working very fast during demonstration but we can hardly do anything on our own. Simultaneously, we do not have similar difficulties with other packages devoted to word processing, data analysis, charting, etc. After a little deeper analysis of the above observations we can put forward the following conclusions, decisive for our further successes or failures in image analysis:

- computers perform very fast *predefined* sequences of operations but are almost useless for development of new, original sequences which are key for any development in image analysis
- we cannot directly use our own experience for development of computerized algorithms for image analysis because we have no detailed knowledge of the functionality of our brain. Moreover, computers are not simplified brains and work in their own, entirely different way

- our visual system is a very efficient tool - we can read nearly unreadable text, we can recognize a person seen only as a distant, walking silhouette or find a proper way from a very simplified plan. But it takes years of training to do it quickly and well. So, no wonder that it is impossible to get, in every case, an immediate, satisfactory solution using computerized image analysis
- stiff, emotionless logic of computers is not good for subtle recognition tasks requiring wide knowledge and intuition
- on the other hand human visual system is sensitive to illusion and, surprisingly, very conservative in its method of analysis. For example, it is very difficult to read a mirrored text
- computers are faster in simple, repeatable operations and therefore offer an ideal platform for measurements.

It is evident from the above analysis that there are diverse, evident and important differences between the human visual system and the properties of computerized image analysis systems. The development of efficient algorithms has to take a lot of time, therefore image analysis should be applied mainly in the case of repeatable tasks, like quality control or scientific research. Recent, computerized technological progress requires numbers - any quantity should be described as 10, 50 or 150 instead of bad, good or excellent. Image analysis systems seem to be an ideal aid for such data treatment.

The variety of material structures being analyzed in industrial and scientific laboratories means that nearly every user of image analysis equipment has to develop from time to time his own, unique procedure. There is only very limited opportunity to use specialists in computer science or interdisciplinary teams for this purpose. Finding the proper solution requires an extremely deep understanding of the processes under analysis and years of experience cannot be summarized within minutes or even hours. Similarly, explanation and understanding of isolated filters used in image analysis, available in numerous textbooks, is insufficient for construction of effective algorithms. The aim of this book is to fill the gap between the theory of image analysis and the practice of material microstructure inspection.

1.2 General concept of the book

The goal, described generally in the previous section, is very difficult to obtain. This difficulty lies in the fact that we need to join two entirely different intellectual spheres: a very strict and highly abstract theory of numerical transformations and often unpredictable, highly practical knowledge of material characteristics.

It seems that the proper solution can be found using some simple rules, briefly described below. First, we will use the terminology common for materials science. Thus we will use *microstructure*, *grain* or *particle* instead of *scene*, *set*, *figure* or *object*. Second, we will avoid mathematical formalism whenever possible. For practical reasons it is less important if the transformation is idempotent, isotropic, homotopic or additive. On the other hand, it is of highest priority to know if the given filter can properly distinguish precipitates of various phases. Third, we will concentrate on typical problems and the simplest solutions, as it seems to be better to tell everything about something than to tell something about everything.

In order to adapt to the needs of various groups of potential readers, the contents of this work are divided into smaller, possibly independent parts. As a consequence, the book is organized into seven chapters, including this one. Their contents are roughly presented below:

- *Chapter one* is devoted to general *introduction* and you are reading it now
- *Chapter two* describes the *main tools for image treatment* and can be recognized as the essence of the transformations most frequently used in image analysis. In other words, this chapter gives the bricks necessary to build an image analysis process. It contains comprehensive descriptions of the nomenclature and basic properties of the transformations as well as some guidelines about where the given family of operations can be successfully applied
- *Chapter three* deals with the problem of *image acquisition and its quality*. Nearly all the transformations, even the simplest ones, of the image are connected to some data loss. Therefore the quality of initial images is of the highest importance. This chapter gives basic rules for specimen preparation, image acquisition and removal of the most frequently met distortions
- *Chapter four* is devoted to *detection of basic features* in the materials microstructure like grains, fibers, pores, etc. These features are essential for understanding of the material microstructure.

However, they often are quite difficult to extract from the initial image

- *Chapter five* covers *treatment of complex structures*, being much more difficult to detect than the basic features described in the previous chapter. Fine and textures structures are analyzed here. The algorithms discussed in this chapter are usually very complex and require understanding of the items analyzed in Chapters two and four
- *Chapter six* gives *analysis and interpretation* of pre-treated images. This chapter describes the technique of digital measurements and their application in microstructural characterization. It discusses properties and specific errors met in digital measurements. Chapter six is somewhat related, discussing basic rules of stereology
- *Chapter seven* summarizes the knowledge previous chapters. Thus it is devoted to *applications and case histories* analysis. The examples discussed in this chapter are selected to show how to solve image analysis tasks, which should be of great value for the novices. Simultaneously, it allows experienced users to confront their own practice with the other viewpoints.

Obviously, the algorithms presented in this book are not exclusive. One can easily find other ways leading to identical or very similar results - this is a very common situation in image analysis. It may also happen that some methods supplied here can be significantly accelerated or simplified. Moreover, the book covers only a small portion of possible tasks and obviously a limited subset of existing procedures is used. These limitations are introduced consciously, in order to keep the volume of the whole work relatively small and to avoid very narrow applications. Once more, the goal of the whole work is twofold:

- to give *effective* solutions to the most common problems met in the analysis of images in materials science
- to *show the way* to reach this effective solution in order to teach the reader to solve his own, unique problems by him- or herself.

Chapter two

Main tools for image treatment

2.1 About this chapter

This book is designed primarily for materials science professionals interested in the application of image analysis tools in their research work. It is assumed that they:

- have a thorough knowledge of materials science as well as wanting to apply image analysis tools quickly and efficiently in their work
- have little experience (if any) with computer-aided image analysis and have no time for in-depth studies of computer algorithms.

There is a subtle dilemma about how to present the image analysis tools for this audience. There is a temptation to offer a general but reader-friendly treatment of computer tools but it would be just another general purpose book on image analysis, of which one can find hundreds on book shelves. Another solution could be to skip all introductory information on image analysis and focus only on specialized algorithms suitable for materials science. However, this also seems to be the wrong approach; such a work would probably be understandable only by a narrow group of specialists, knowing all the tips prior to reading this book. To make things more complex, the text should not refer to any existing software. Thus, the lack of standardized nomenclature should be also taken into account.

The solution chosen here is to give a general description of all the main groups of transformations, without any reference to detailed analysis of the algorithms, formal restrictions, etc. This information is divided into two independent but complementary parts for all the groups analyzed:

- verbal description, giving the general properties of the transformation analyzed as well as the possible application directions
- graphical illustration, showing the sample image before and after the transformation.

Verbal description covers a general idea of the transformation analyzed, together with its potential application area. Furthermore, to enable it to work more easily with a great variety of software, the most commonly used synonyms are cited. It is very significant that no formal definition of the procedure body or parameters, nor analysis of the algorithms available, are submitted. The aim of this introductory part is to provide the very basic knowledge necessary for individual work with image analysis packages. The reader should learn what is possible from a given family of transformations. He should also possess at least some rough knowledge concerning possible application areas. Afterwards, detailed data on image analysis algorithms can be found in specialized literature or software documentation.

Graphical illustration covers both initial and post-processing images, thus enabling one to get the feeling of what *direction* in image alteration can be expected for a given family of transformations. To allow one to compare various transformation families, the same sample image is used whether possible. Additionally, the line profile (plot of pixel values along a line) at exactly the same location is added to all the images. This allows a more quantitative way of exploring the changes in image data.

2.2 *Basic image enhancement*

Any image discussed here is a mosaic of very small areas, called pixels, filled with a single gray level or digitally defined color. Thousands of pixels, touching each other and placed within a (usually square) grid, give us the illusion of a realistic, smooth picture. This pixel nature of computerized images allows us to store them and transform them as matrices of numbers. This is the very basis of computer-aided image analysis.

Gray images are usually described by 256 gray levels. This corresponds to 8 bits per pixel as $256 = 2^8$. In this representation 0 equals black and 255 denotes white. 256 gray levels are quite sufficient for most applications as humans can distinguish approximately only 30 to 40 gray levels. In some applications, however, other depths of image data are used: 2 (binary images), 12, 16 or 32 bits per pixel.

Color images are most commonly stored as RGB (Red Green Blue) images. In fact, each of the RGB channels is a single gray image. Analysis of color images can be interpreted as the individual analysis of the gray components put together at the end to produce the final color image. Thus, understanding the principles of gray image analysis gives sufficient background for color image treatment.

Due to the digital nature of the computer images described above they can be modified using usual mathematical functions. The simplest functions can be applied for basic image enhancement, usually known as brightness and contrast control. Some selected functions of this type are schematically shown in Fig. 2.1. Illustrative examples, as described in Section 2.1, are shown in Figs. 2.2 and 2.3.

In the case of 8 bit images, any transforming function has only 256 values corresponding to 256 argument values. So, instead of defining the function and calculating its value for each pixel, it is much simpler and quicker to define a table of 256 values, which can be very quickly substituted in the computer memory. This method of computation is extremely useful in computers. The tables of pixel values are usually called LUT (Look-Up Table). Thus, instead of defining the transform function we quite often define the LUT.

Brightness and contrast control in image analysis are fully analogous to the brightness and contrast adjustments in any TV set. Increased contrast can cause the loss of some data. Part of the dark gray levels can be converted into black and part of the bright pixels can be converted into white (see Figs. 2.1 and 2.2b). These negative effects can be avoided or significantly reduced after using a suitable combination of both transformations; for example, brightness with lower contrast.

Brightness and contrast are useful for visualization purposes but in general, due to the possible loss of data, are rarely applied in image analysis. There is, however, one exception usually called *normalization*. This is a kind of brightness/contrast modification leading to the image with the lowest pixel values equal to 0 (or black) and the highest pixel values equal to 255 (or white). Usually, if one analyzes a series of images they vary in contrast and brightness. This effect can be caused by numerous factors, like apparatus aging, voltage variation, dust, etc. Normalization allows us to alter these images as if they were recorded in very similar or identical brightness and contrast conditions. Therefore, normalization is quite often applied as the first transformation in image analysis.

In a similar way, one can also produce the *negative* or *inversion* of the image. It is one of the simplest LUT transformations. White becomes black and vice versa. If we add the initial image and its negative, we will get an ideally white surface. The negative can be used for some special purposes, described later in this book.

Due to its non-linear characteristics, the human eye is more sensitive to changes in the brighter part of the gray level spectrum than in the darker one. This can be easily noted in Fig. 2.1, where one can analyze two rectangles with blend fills from black to white. Try to choose the region filled in 50% with black. Most probably you will choose a point which is closer to the black side of the rectangle, whereas 50% black is exactly in the middle. As a consequence of this non-linearity we can notice many more details in the brighter region of the image than in its darker part.

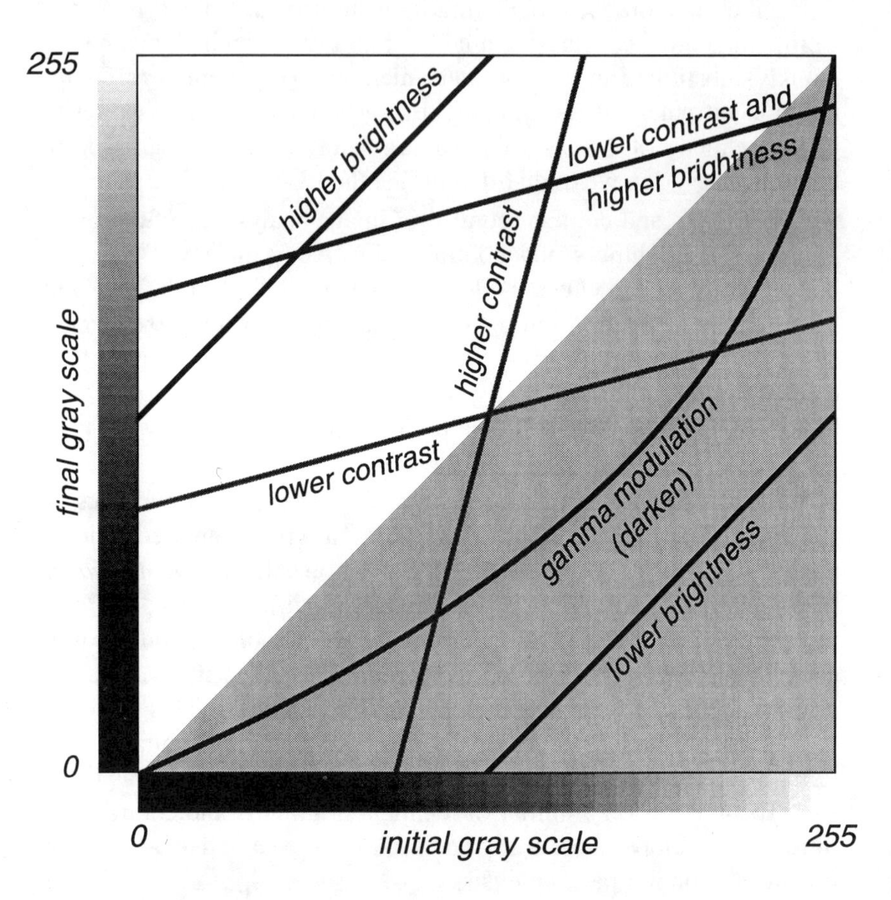

Fig. 2.1. Selected functions for basic image enhancement.

So, to get an image with details easily seen in the whole image, one should stretch the dark and squeeze the bright range of gray levels. This can be done with the help of *gamma modulation* (see Fig. 2.1). An example of this transformation is shown in Fig. 2.2. Note that at first glance the result of gamma modulation seen in Fig. 2.2c is very similar to the image produced by increased brightness (Fig. 2.2b). Closer analysis shows the difference in the brightest areas. A bright particle in the lower right corner is entirely white after increased brightness whereas after gamma modulation all the details are still visible, as in the initial image.

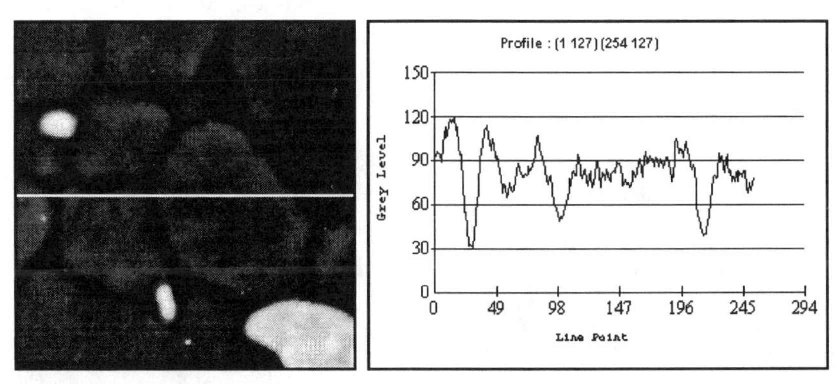

a) initial image

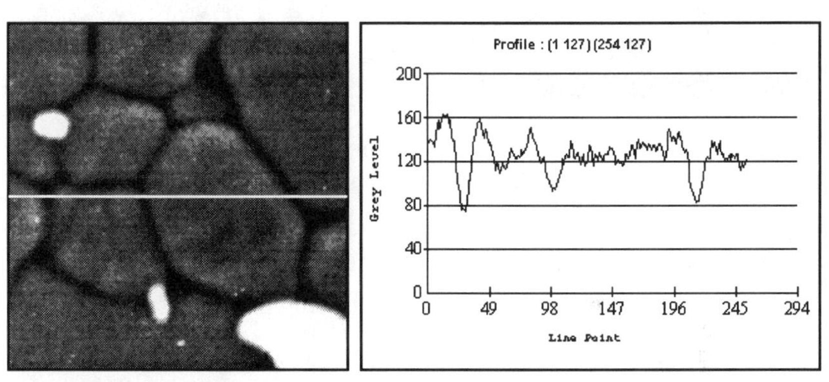

b) initial image with higher brightness

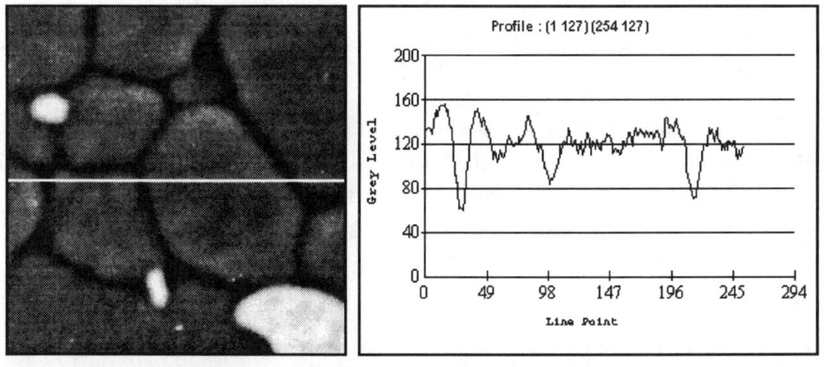

c) initial image after gamma modulation

Fig. 2.2. Brightness and contrast control.

Another example of gamma modulation, applied to a fracture surface, is shown in Fig. 2.3. It should be pointed out, however, that all the details visible after gamma modulation obviously exist in the initial image. This transformation only makes them visible to the human eye.

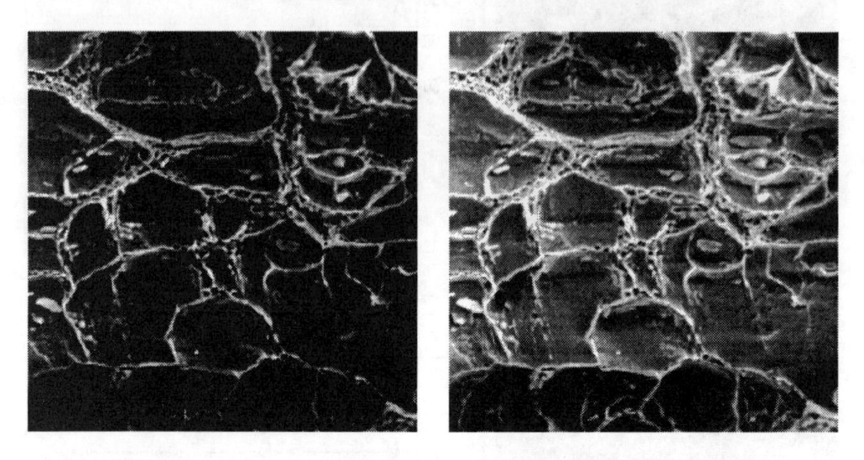

Fig. 2.3. Gamma modulation (right image) allows observation of details in the darker part of the fracture surface (left image).

Another interesting non-linear LUT transformation is known as *histogram equalization*. This has the following properties:

- it preserves the natural sequence of grays, similarly to gamma modulation. In other words, features darker in the initial image remain darker in the transformed image
- if we divide the whole gray scale into small classes of equal size, the same number of pixels will be observed in each class and the histogram of gray levels will be flat (equalized).

Histogram equalization can produce images with somewhat unnatural appearance (see Fig. 2.4b), but simultaneously it produces an image with the highest possible contrast, preserving approximately all the details of the initial image. As will be shown later, histogram equalization is useful for advanced and automatic thresholding (binarization).

There exist many other LUT modifications and they are applied for artistic or visualization purposes. They have much less meaning for extracting features from images, as their results are often unpredictable.

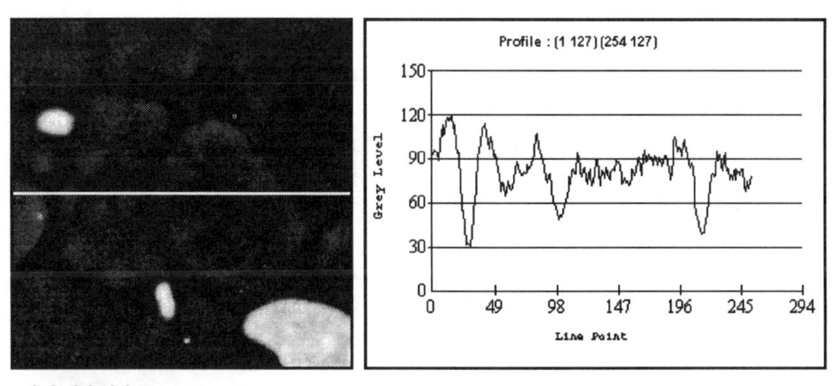

a) initial image

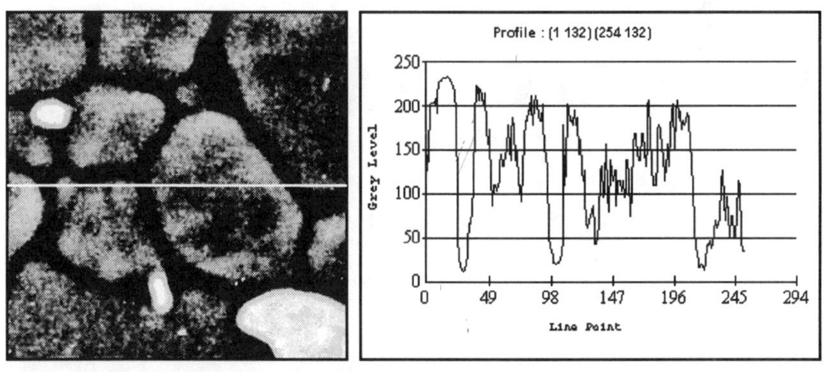

b) initial image with equalized histogram

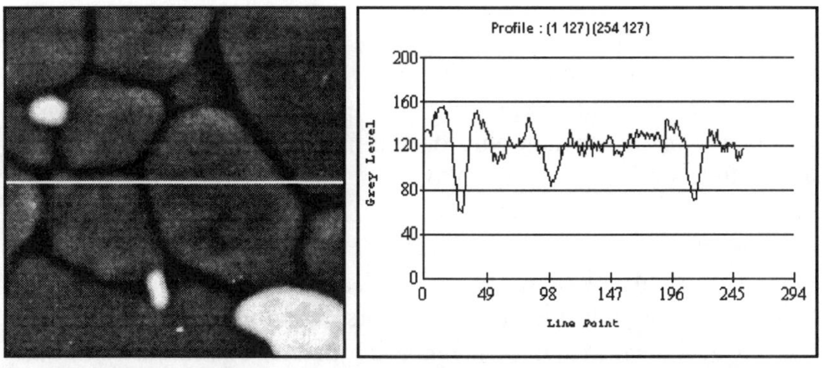

c) initial image after gamma modulation

Fig. 2.4. Effects of histogram equalization and gamma modulation.

2.3 Filters

Filtering is one of the most common processes in nature and technology. One meets filters in everyday life: sand and earth filter polluted water and make it clean, paper filters produce tasty coffee or tea, vacuum cleaners filter dust particles out of the air, electronic filters smooth radio signals which lead to perfect sound or video images, etc. Filters of various types are also among the most frequently used tools for image treatment.[7,13,21,70,80,84,85,87] The principle of filtering is schematically and intuitively shown in Fig. 2.5.

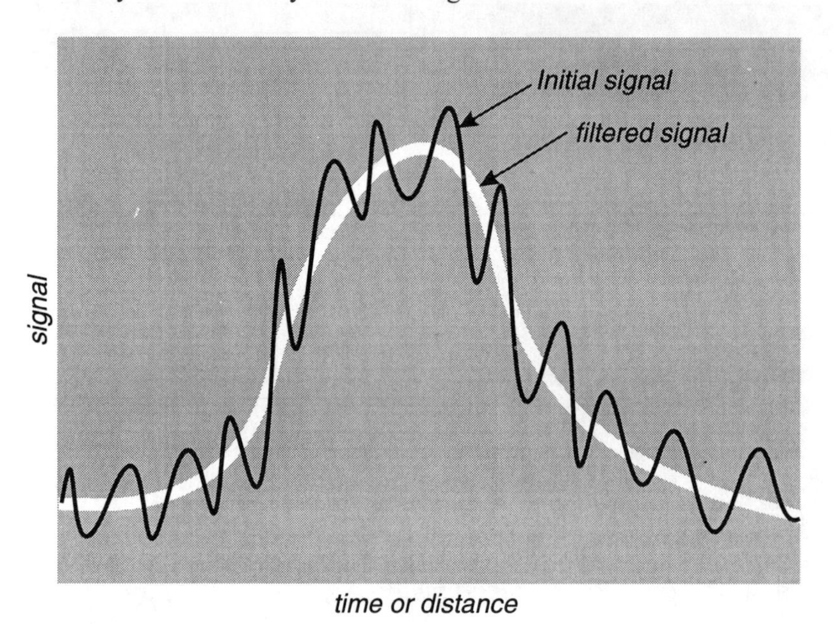

Fig. 2.5. Filtering process (schematically).

The transformations described in Section 2.2 can be called *point-type operations*. This means that the result of any transformation of any image pixel depends only on the initial gray value of this pixel and is independent of its neighbors. For example, the negative of any white point is always black, whatever the gray levels are of the surrounding pixels. By contrast, filters are *neighbor-type operations*. In other words, the pixel value after filtering is a function of its own value and the gray levels of its neighbors. Usually, filters return values that are weighted means of neighboring pixels. The majority of software packages offer numerous predefined filters as well as user-defined ones. In this last case the user can define the matrix of coefficients used to compute the weighted mean returned by a filter.

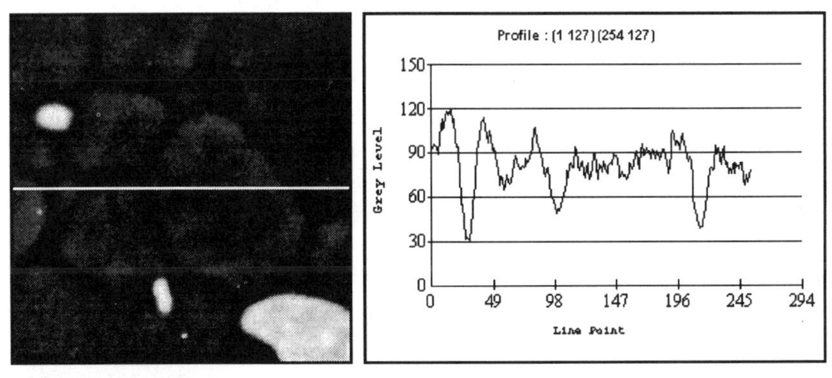

a) initial image

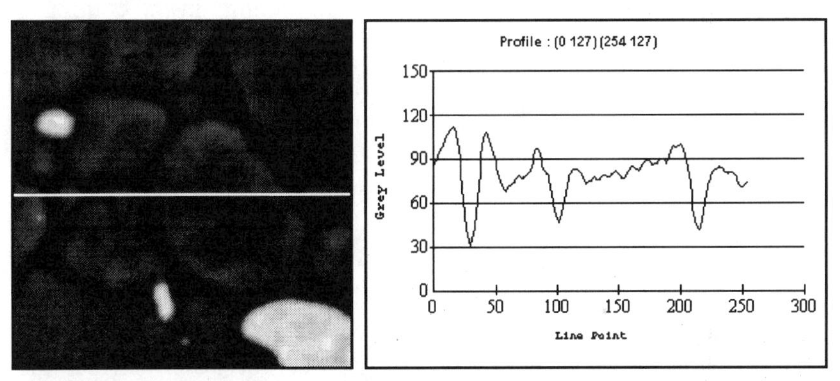

b) initial image after smoothing filtering

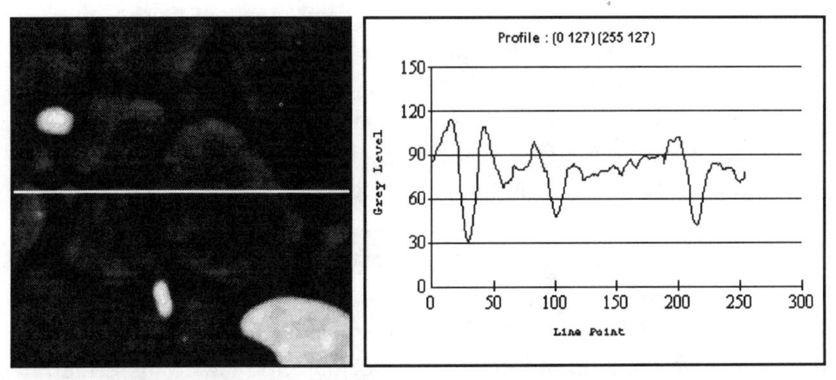

c) initial image after median filtering

Fig. 2.6. Simple filters for noise reduction - smoothing and median.

Digital images are often polluted with noise produced, for example, by video cameras in the case of insufficient illumination or by SEM detectors. Obviously, noise should be removed from such images prior to any quantitative analysis. This can be done using suitable filters.

In Fig. 2.6 one can analyze the effect of two simple filters suitable for noise reduction. This effect is rather subtle, so it is more visible in the profile plots than in the images. The first, *smoothing filter* (Fig. 2.6b), is probably the simplest possible filter - it returns just an arithmetic mean of the pixels in a 5x5 size square. Matrices of sizes 3x3 and 7x7 are also very common. In more advanced packages, larger matrices are available as well. They can be called a *box filter,* an *averaging filter*, etc. The smoothing filter provides an image with reduced noise and a somewhat out-of-focus appearance. To reduce this last phenomenon, other filters, for example, *Gaussian*, are introduced. In these filters, diverse points have different weights in the computed average; generally the weight is smaller for pixels more distant from the central (just altered by a filter) pixel.

Smoothing filters work well if the image is not excessively noisy. In other cases they produce unsuitable results. Let us analyze it with an example. Assume we use a 3x3 kernel and the pixels have the following values put in ascending order:

6, 8, 12, 15, 15, 17, 19, 20, 95.

It is evident that the pixel values are in the range from 6 to 20 and the pixel with value of 95 should be thrown away. If we compute the arithmetic mean value, as a smoothing filter does, we will get the value of 23. Simultaneously, the arithmetic mean from the first eight values is equal to 14. This last value is both intuitively acceptable and far from 23. So, in this case a smoothing filter does not work well.

Better results can be obtained if we use a *median filter* (see Fig. 2.6c). The median is the value situated exactly in the middle of the series of numbers set in ascending order. In the example analyzed above, it would be the fifth value, i.e., 15. So, a median filter can be effectively applied for treating heavily noisy images and in most cases is the best solution available. Moreover, this filter has two important properties: it does not add new values to the image data (median is one of the already existing values) and it keeps the image sharp.

Noise, especially of a periodic character, can also be efficiently removed with the help of Fourier transformations. This transformation is, however, much more difficult to perform. There are some restrictions to the images and only advanced packages offer efficient tools for Fourier analysis. It will be described in Section 2.6.

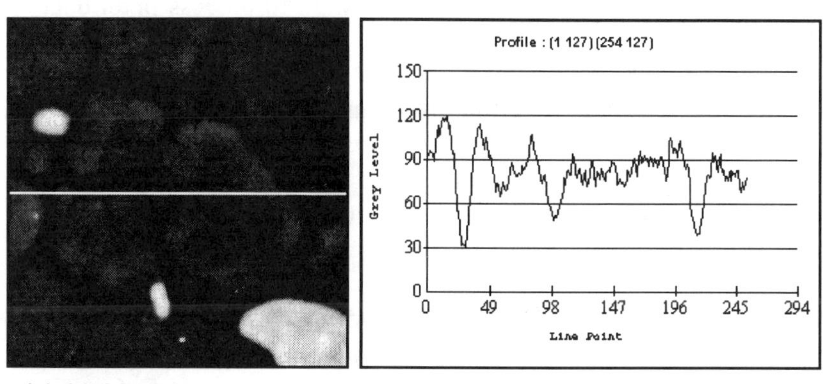

a) initial image

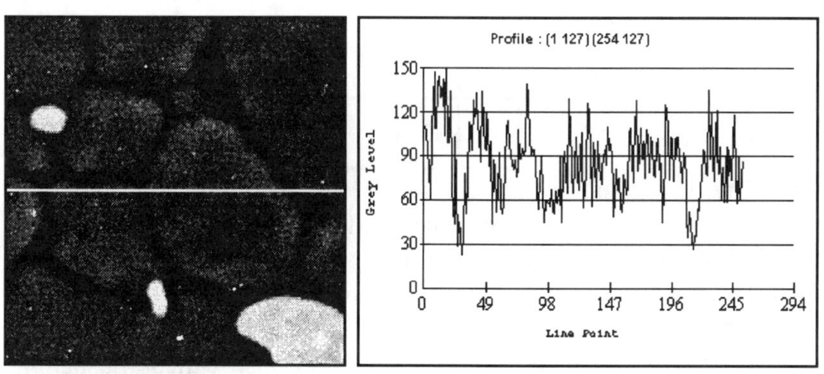

b) initial image after sharpening filtering

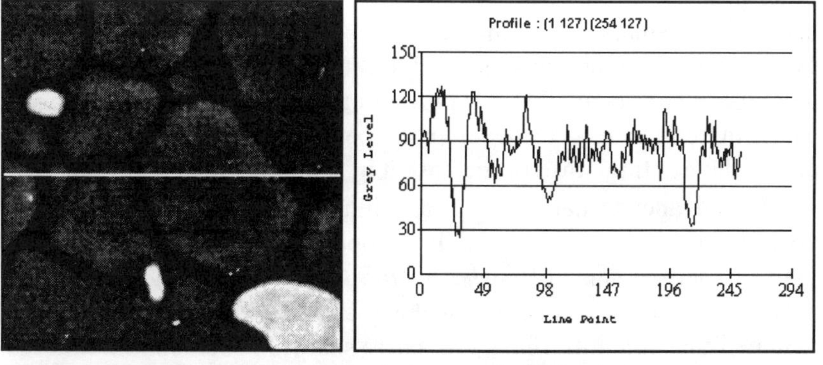

c) initial image after advanced sharpening filtering

Fig. 2.7. Examples of sharpen filters. Note that the side effect of sharpening is an increase in noise level (see plots).

Noise (see plots in Figs. 2.6 and 2.7) is a local feature, generating narrow peaks in gray level plots. If we want to process pixel data as signals, noise is recognized as a high frequency part of the signal spectrum. In order to remove noise, one should filter out the high frequency part and pass only the low frequency component. Therefore, smoothing filters are often called *low-pass filters*.

Obviously, there exist filters with just the opposite properties, called *high-pass filters*. These filters strengthen the high frequency component of the picture data. In principle, they are also weighted means of pixel data, properly designed in order to increase the contrast locally (difference between neighboring pixels). A typical example of such a filter of size 3x3 can be shown in the form of the following kernel:

-1	0	-1
0	5	0
-1	0	-1

This filter works in the following way: if all the pixels have the same values, the new value remains unaltered as the sum of the weighting coefficients equals 1. If the central point has a value two times greater than its neighbors, its value after the transformation will be six times greater (2*5-4*[-1]=6). Any image treated in this way looks sharper (see Fig. 2.7b) and therefore such filters are also called *sharpening filters*. Of course, one can design hundreds of sharpening filters and everyone can experiment with them using *user-defined filters*. The basic property of sharpening filters is that the sum of the weighting coefficients equals 1. The image after applying such filtering is, however, noisier (see the plots situated to the right of the images). Generally, the increase in noise is proportional to the sharpening effect. Thus, there is a need to develop sharpening filters which do not increase the existing noise or increase it by a limited degree.

An advanced sharpening filter, introducing a small amount of noise is presented in Fig. 2.7c. It is popular in photo-retouching programs under the name of *unsharp mask filter*. Its principles have their roots in advanced retouching techniques used in photography. It is assumed that the initial image is fairly sharp. In such conditions it is possible to detect the edges as a difference between the original and smoothed images. Edges extracted in this way are subsequently added to the initial image, thus producing a sharpened picture without unnecessary noise (to observe this effect consider the plots in Fig. 2.7).

Unfortunately, we have no tools for sharpening the image completely without adding some noise or losing some pixel data.

Sharpening filters are widely used in typography. In practice, all printed illustrations are electronically sharpened before sending them to the printing press. This produces nice looking images and due to the properties of human eyes, the existing noise is not disturbing. By contrast, the aim of image analysis is not to enhance images but to extract some features or information from the image. In such circumstances the noise accompanying the sharpening process is very annoying and can even make further analysis impossible. Therefore, sharpening filters are rarely used in image analysis. The only exception is an edge detection process, described later in this chapter in more detail.

The number of possible filters is fairly unlimited and some guidelines on how to design them and an analysis of existing algorithms can be found in the specialized literature. In this short description we will present only the main properties of selected filters in order to give the reader some feeling of these properties. There are, however, two important filters worth describing in more detail. The *minimum* and *maximum filters* are widely used in practice and simultaneously are equivalent to some morphological operations, which are sometimes decisive for the final result of image analysis.

The *maximum filter* (Fig. 2.8b) returns the value which is equal to the maximum of all the pixels surrounding the pixel being analyzed. As a consequence, one obtains a new image which is brighter than the original, with removed noise. The filtered image contains less details than the initial one. In such a filtered image it is easier to detect large-scale features, like, for example, grains.

The *minimum filter* (Fig. 2.8c) is just the opposite transformation. It returns the value which is equal to the minimum of all the pixels surrounding the pixel being analyzed. It can be also interpreted as a maximum filter of the negative of the initial image. The result of minimum filtering is darker than the original and contains less details. A combination of these two filters (maximum and minimum) gives a new filter, suitable for noise filtering.

Obviously, the proper use of filters requires some experience and similar results often can be achieved after entirely different sequences of operations. This short description should show you that the filtering principles are not as difficult as they look at first glance and they help you to navigate among various filters and experiment with their application.

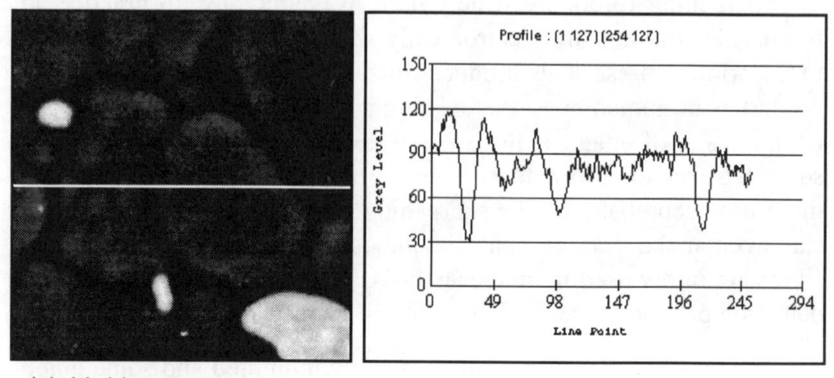

a) initial image

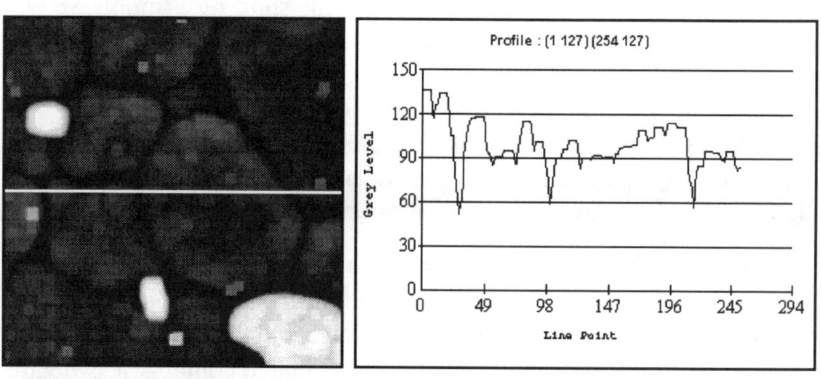

b) initial image after maximum filtering

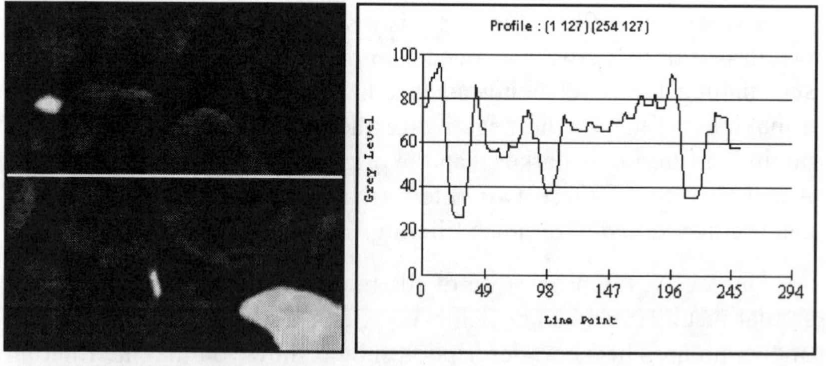

c) initial image after minimum filtering

Fig. 2.8. Examples of maximum and minimum filtering.

2.4 Binarization

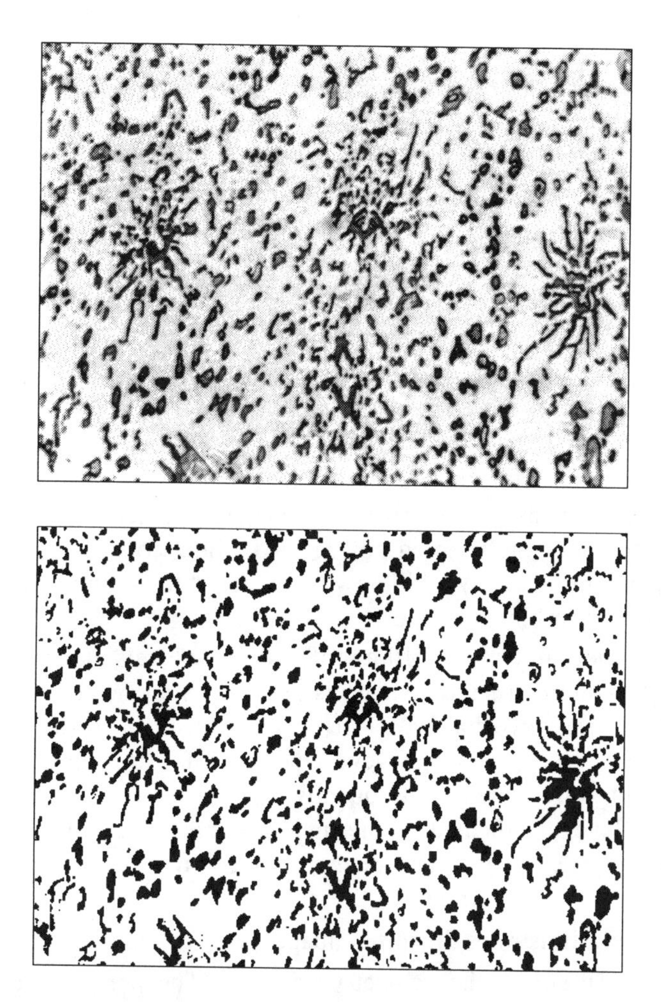

Fig. 2.9. Gray (top) and the corresponding binary (bottom) image.

Currently, gray images are the most frequently used aid for recording image data in materials science. Images presented in Sections 2.2 and 2.3 are stored just in gray levels. However, if we go back for a moment to history, the first automatic image analyzers worked only on binary images,[45] i.e., images made out of black and white points. Even now, binary images (Fig. 2.9) are commonly used in image analysis. There are at least three important reasons for application of binary images:

- binary images allow one to save a lot of memory, as the appropriate files are approximately 8 times shorter than in the case of gray images and 24 times shorter than in the case of full-color images
- only in binary images can one detect separate features, for example, particles or grains (every connected set of pixels is recognized as a single particle). Consequently, binary images are necessary for counting objects and for such measurements as area, perimeter, diameter, deviation moments, location of a center of gravity, etc.
- some transformations, mainly from the family of morphological operations, can be performed only on binary images. There are, for example, procedures for separation of particles glued together, lost grain boundary restoration and some simulations.

The process of transformation of gray-scale images into binary ones is called *binarization* or *thresholding*. Its principles are illustrated in Figs. 2.10 and 2.11, respectively. In Fig. 2.10 one can observe a microstructure of a sintered steel. Three main structural constituents can be easily recognized in this picture: black pores, gray and convex grains of the sintered powder and white, concave precipitates of the bonding phase. We will now try to detect these constituents and to generate their binary images.

At the bottom of Fig. 2.10 a profile of gray levels along the white line in the microstructural image is shown. One can note from this profile that gray levels above 215 (threshold level A) correspond with the bonding phase, and gray levels below 130 (threshold level B) correspond to the pores. Consequently, gray levels between these two threshold levels (indicated by the light gray belt in the plot) coincide with the powder phase. The results of the binarization process described above are shown in Fig. 2.11:

- Fig. 2.11a illustrates the initial image
- Fig. 2.11b shows the geometry of pores. The pores are detected from the initial image as all the pixels with gray levels *below* the threshold level B (Fig. 2.10). Thus, this kind of thresholding is sometimes called *binarization with an upper threshold*
- the bonding phase, shown in Fig. 2.11c, is detected as all the pixels with gray levels *above* the threshold level A (Fig. 2.10). Consequently, this kind of thresholding is sometimes called *binarization with a lower threshold*
- the remaining powder grains are detected with the help from two, upper (A) and lower (B) thresholds. Such a type of operation can be called *dual threshold binarization* and is very often met in practical applications.

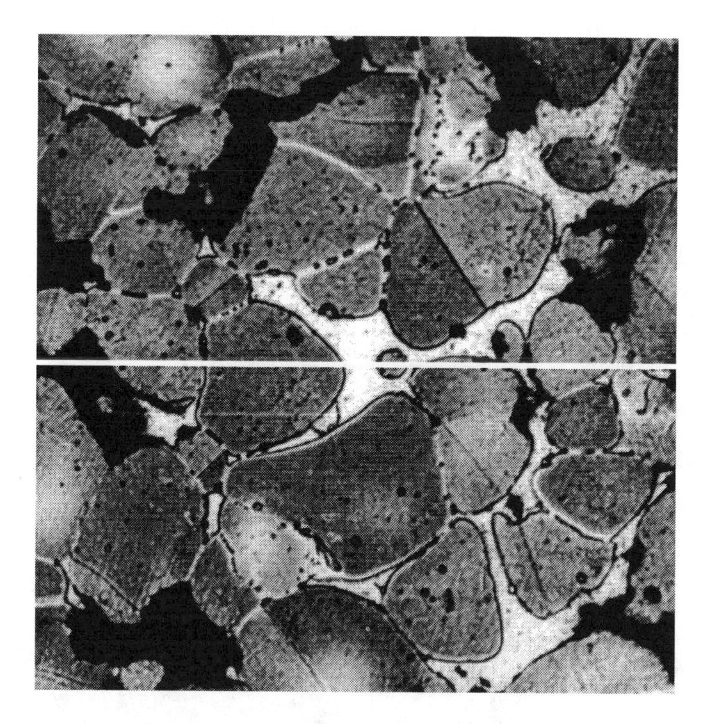

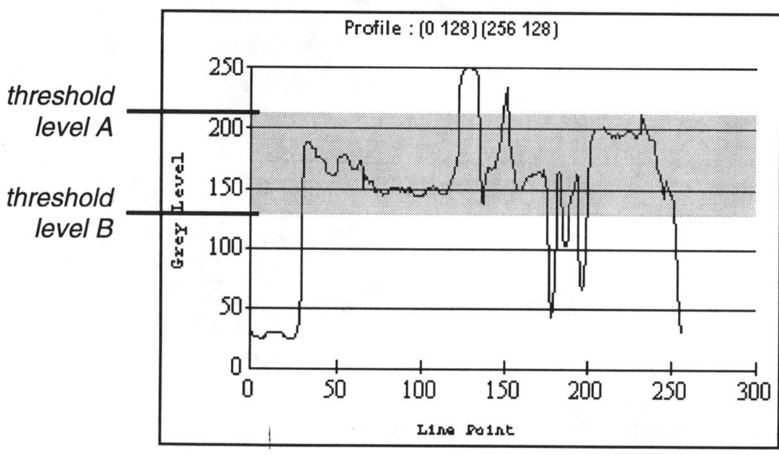

Fig. 2.10. Exemplary microstructure and threshold levels suitable for binarization of structure constituents.

It is clear from the example demonstrated above that the proper choice of threshold level is decisive for the results of analysis. If we have two distinct phases with two different gray levels, the appropriate threshold can be relatively safely chosen as an arithmetic mean of these gray levels. In the case of an irregular gray-level distribution (see plot in Fig. 2.10), proper choice of the threshold level is much more complicated and sometimes even impossible. In such cases adequate treatment has to be done prior to binarization - a description of such cases constitutes the core of this work.

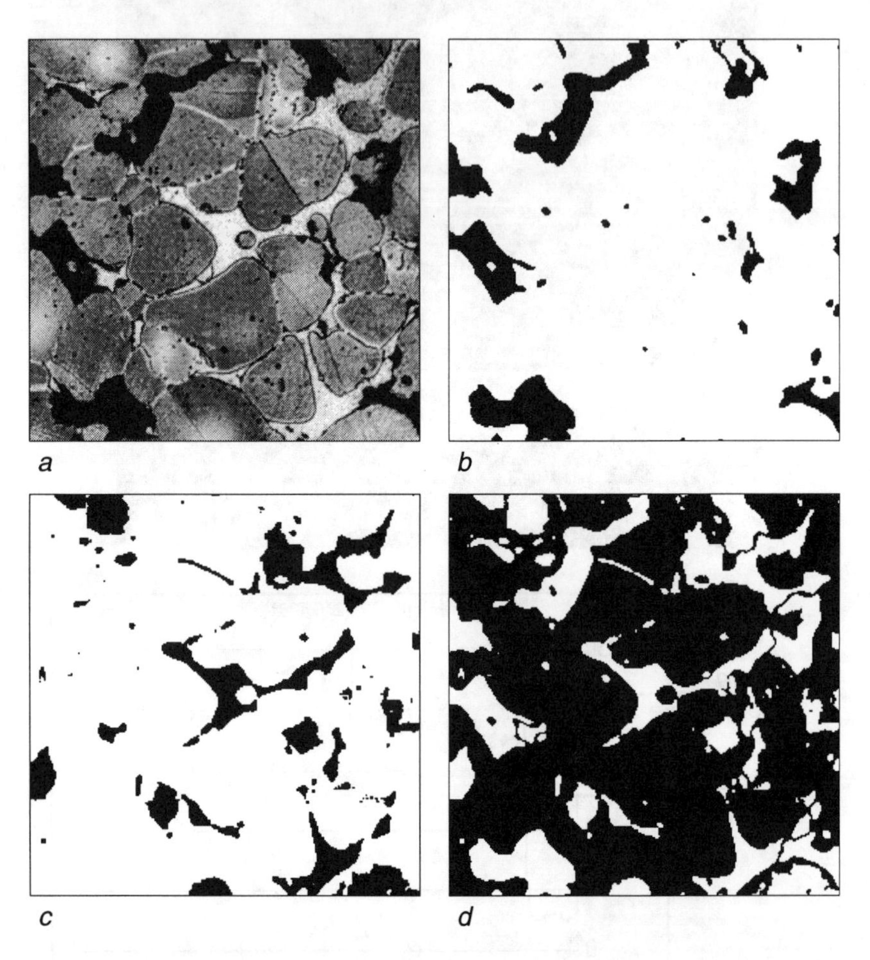

Fig. 2.11. Initial gray-scale image (a) and corresponding binary images of pores (b), bonding phase (c) and powder phase (d).

In practical applications it is usually advisable to use interactive binarization. In other words, one should choose the threshold level, judge the result and, if necessary, correct this level. Such a solution works quite well but is sensitive to human error and, what is more important, cannot be applied in the case of fully automatic analysis of a huge number of images. In such cases one can try to use *automatic thresholding*. The idea of automatic thresholding is demonstrated in Fig. 2.13. Let us assume we have an image with two phases to separate and the total amount of the phase to be detected is relatively small. In such circumstances one can watch a gray-level distribution similar to the plot in Fig. 2.12. An appropriate threshold level can then be determined automatically on the basis of the gray-level distribution. In the case of a bimodal distribution the threshold will correspond to the local minimum, lying between two local maxima (Fig. 2.12). Such an automatically determined threshold can give identical detection results, irrespective of the image contrast and brightness, as shown in Fig. 2.13.

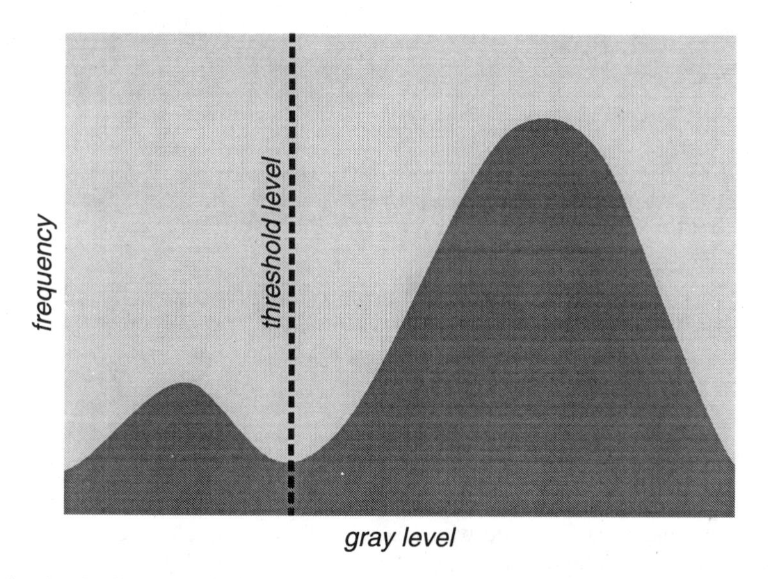

Fig. 2.12. Gray-level distribution function and the corresponding automatic threshold level.

The appropriate threshold can also be determined in other ways. In many cases (composite materials, hard sinters, cast iron, etc.), one can compute precisely the volume fraction of the main structural constituents. For example, in the case of gray ferritic cast iron the volume fraction of graphite can be determined from the equation:[106]

$$V_V = 0.034\%C + 0.005 \quad (2.1)$$

It is well known from basic stereological relations[100] that

$$V_V = A_A = L_L = P_P \quad (2.2)$$

Thus, knowing from the chemical composition that our cast iron has 4.3%C, one can easily compute the corresponding area fraction of graphite as approximately 0.15 or 15%. On an as-polished specimen one will see only dark graphite precipitates and a bright matrix. After histogram equalization we can be sure that graphite will occupy the lowest 15% of the gray level spectrum. If 0 denotes gray and 255 white, respectively, the appropriate threshold level can be fixed at a value of 38.

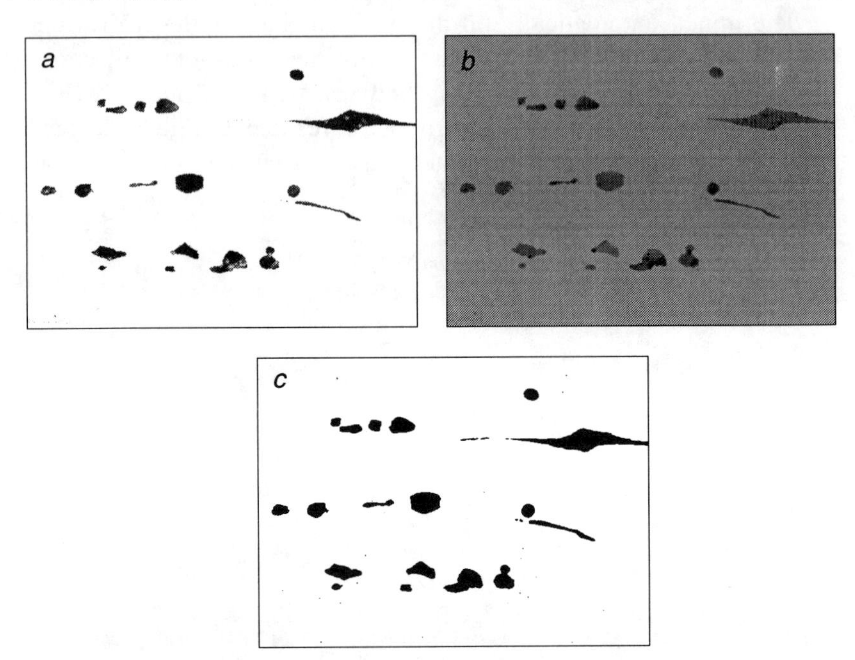

Fig. 2.13. Two gray-scale images (a and b) containing the same particles but with entirely different contrast and brightness levels together with the resulting binary image (c), obtained with the help of an automatic threshold level. The detected binary image is identical for both images (a) and (b).

Many researchers work on new, more flexible binarization procedures. Automatic thresholding, as discussed above, is one of the promising results. Another type of threshold processes is a conditional procedure called a *hysteresis threshold*. It can be easily replaced by a series of simple operations, but for higher clarity and computation speed it is very convenient to have such a tool available. Its properties are interesting for practical applications and, therefore, the hysteresis threshold method will be discussed in some more detail.

Let us analyze an image shown schematically in Fig. 2.14. One can see in this image two objects, A and B, having the same gray level. The difference between them is that object B contains two very bright spots. Consequently, on the gray-level distribution plot one can fix two threshold values: the basic threshold will detect both objects, including bright spots (Fig. 2.14b) and the marker threshold will detect only the spots. In such circumstances the hysteresis threshold can be defined as a procedure for the detection of objects according to the basic threshold level *under conditions* that any markers, detected by the marker threshold, belong to the detected objects (Fig. 2.14c).

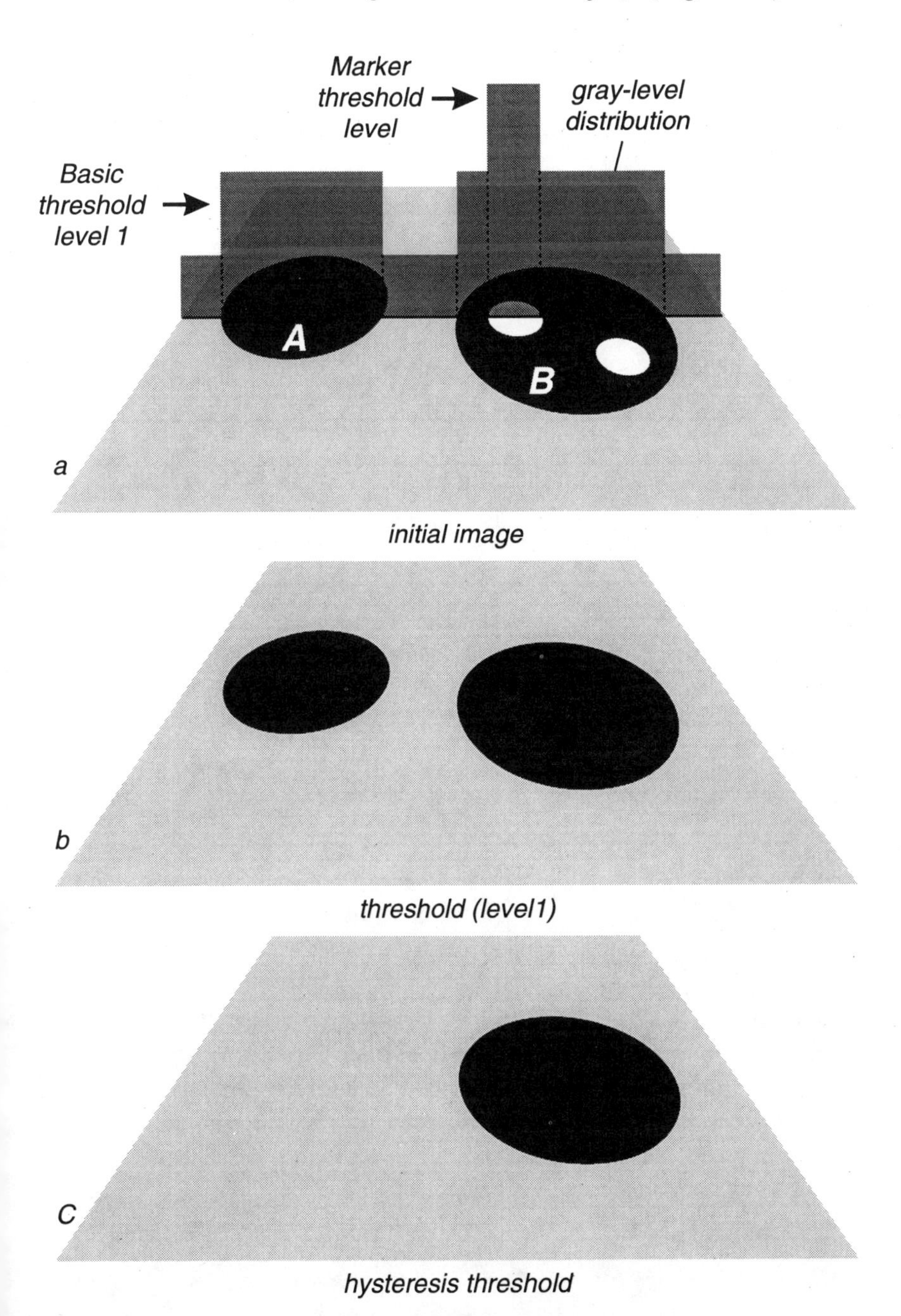

Fig. 2.14. Schematic illustration of the hysteresis threshold.

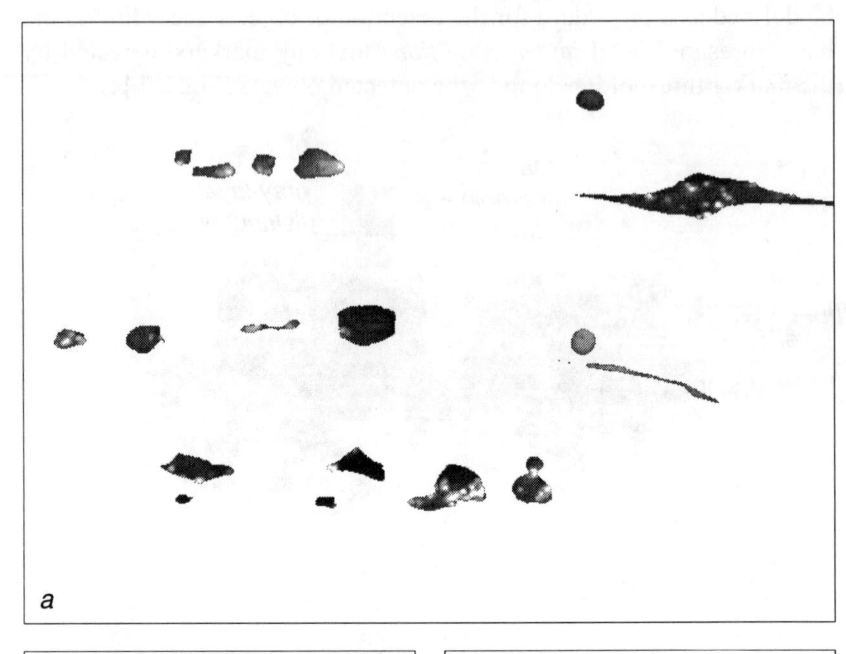

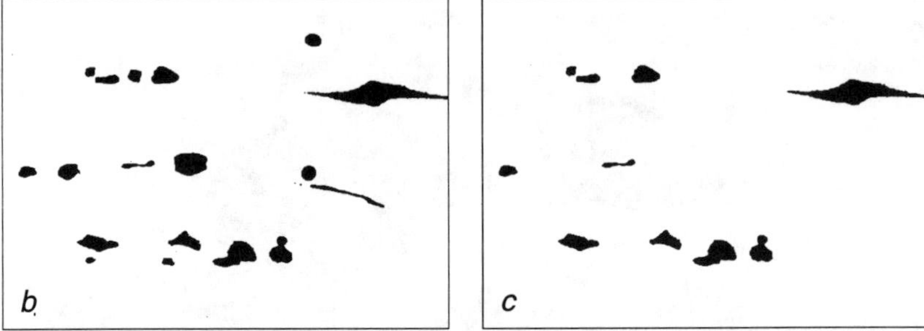

Fig. 2.15. Initial image (a) and the same image after classical binarization (b) or hysteresis threshold (c).

An example of the application of the hysteresis threshold method is shown in Fig. 2.15. The particles visible in this picture are non-metallic inclusions. In the original image, bright regions in the inclusions are their darkest parts. In order to apply hysteresis threshold a negative of the inclusions was used.

2.5 *Mathematical morphology*

Mathematical morphology[21,80,84,85,114] is a highly abstract theory of image transformations, possessing its own rules and notation. Due to their complexity, morphological operators are implemented only in advanced packages. On the other hand, mathematical morphology enables detection of various features in the image in a way somewhat similar to human intuition. As a consequence, application of morphological operators enables detection of features not available with other analysis methods. Therefore, without any description of the mathematical formalism of morphological transformations, we will describe the basic concepts and their application.

From the very long list of morphological transformations available, the following groups will be shown in more detail:

- hit or miss transform
- erosion, dilation, opening and closing
- skeletonization and SKIZ
- watershed detection.

The central point of mathematical morphology is the concept of *structuring element*. It can be understood as a model of local pixel configuration. Usually, structuring elements are defined using the following notation:

1 - for pixels belonging to the set of points under analysis (for example, black points in Fig. 2.15b, c)
0 - for pixels belonging to the matrix (for example, white points in Fig. 2.15b, c)
X - for pixels not taken into account (i.e., this point can have any value and has no effect in the transformation).

This definition is suitable for binary (black and white) images. For gray images the meaning of the symbols defining a structuring element changes slightly: '0' denotes pixels darker than the given pixel while '1' denotes pixels lighter than the given pixel.

An exemplary structuring element is shown below:

0	0	0
0	1	0
0	0	0

It is easy to guess that this element illustrates an isolated pixel - a single pixel surrounded by the matrix. The example above refers to the most common case of a square grid of pixels. However, there exists also another solution, based on a hexagonal grid of pixels. This last solution has some advantages in the analysis of fine, highly curved features, but more detailed analysis of the relation between these two types of grids exceeds the goals of this book. Suitable information can be found in monographs devoted to mathematical morphology and the theoretical background of image analysis.[21,84,85] Here we will only show how the structuring element for isolated pixel looks in the hexagonal grid:

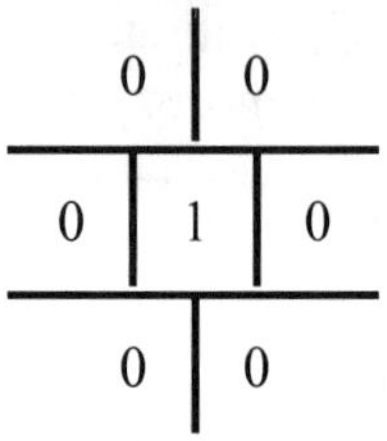

0 | 0
0 | 1 | 0
0 | 0

It is quite difficult to feel the subtle differences between morphological operations and other tools used in image analysis. With accuracy sufficient for this work, we can define morphological transformations as advanced filters, applied not for all the pixels in the image, but only for pixels that fit configurations defined by the structuring element.

The *Hit or miss transform* (HMT) can be recognized as the most general morphological operation. HMT removes all the pixels that do not fit with a configuration defined by the structuring element. So, applying HMT to the element shown on the previous page we will detect isolated points. Using the following element:

1	1	1
1	1	1
1	1	1

HMT will preserve all the internal points (surrounded by '1') and remove all the points touching the matrix (at least one '0' in the closest neighborhood). Such a transformation is known as *erosion* and can be defined in a completely different way, which will be shown later in this chapter.

One can also introduce HMT with a *rotating structuring element*, i.e., a sequence of HMTs with different structuring elements, obtained by rotation of the initial configuration. This concept allows for new properties of the whole transformation. Let us take the following element:

X	0	X
X	1	X
X	X	X

Rotation of the above configuration will produce a sequence of eight structuring elements:

X	0	X
X	1	X
X	X	X

X	X	0
X	1	X
X	X	X

X	X	X
X	1	0
X	X	X

X	X	X
X	1	X
X	X	0

X	X	X
X	1	X
X	0	X

X	X	X
X	1	X
0	X	X

X	X	X
0	1	X
X	X	X

0	X	X
X	1	X
X	X	X

After this transformation, we will detect all the boundary points, i.e., points touching the matrix. This is done in the following way: at least one of the surrounding points is equal to 0, which is equivalent to belonging to the particle boundary. The rest can be '0' or '1', as in the structuring element we have 'X'. This ensures a fit to any local boundary configuration. Rotation enables fitting to all the possible configurations.

The above examples explain the basic properties of HMT, obviously without exploring the whole problem. However, this knowledge should be sufficient for individual work and experiments with this transformation if implemented in the package. If there is a possibility of switching between square and hexagonal grids, try to explore both of them and note the differences.

Now we will discuss a group of four transformations, known as *erosion, dilation, opening* and *closing*. These transformations can be successfully defined using the concept of HMT, but other definitions, presented below, seem to be more instructive.

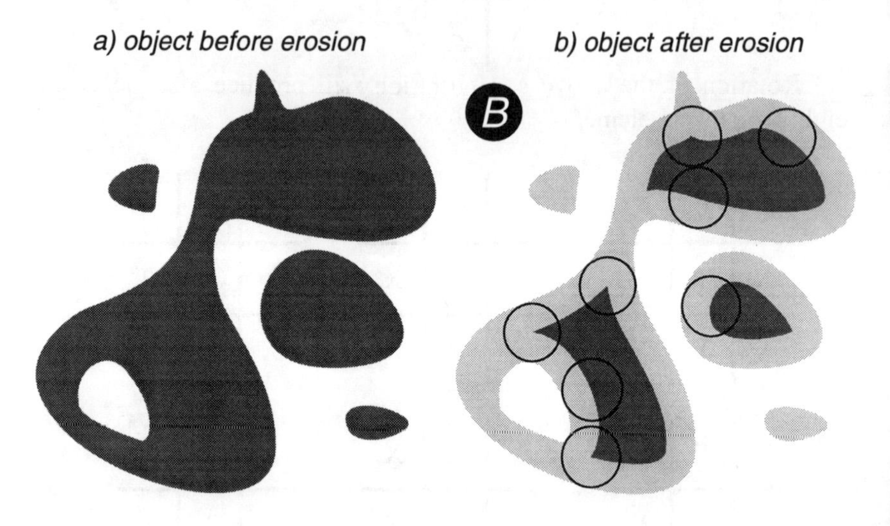

Fig. 2.16. The principle of erosion.

Let us take into consideration a figure as drawn in Fig. 2.16a and take a circular structuring element ***B***. Now, let us roll element ***B*** on the internal side of the particle boundary (Fig. 2.16b). The subsequent positions of the circle center will create the eroded object (darker figure in Fig. 2.16b.

The above definition is absolutely equal to the above-mentioned erosion by HMT. There also exists a third variant: erosion that can be recognized as a minimum filter. Indeed, if erosion removes all the pixels touching the matrix (points with value of 0), changing the pixel value to 0 (the minimum in the local neighborhood) produces exactly the same effect.

The objects in Fig. 2.16 reveal all the basic properties of erosion:

- eroded object is smaller than the initial one
- some narrow peninsulas or small particles will disappear after erosion
- erosion can divide a coherent object into some smaller and isolated objects whose shapes are totally different from the initial object

- erosion is additive, i.e., two erosions with structuring element ***B*** are equal to an erosion with the element ***2B***, two times larger.

Dilation is a transformation just the opposite to erosion. It can be defined as a maximum filter or produced by a circle as in Fig. 2.16, but rolled on the external side of the initial figure. Last, it can be interpreted as erosion of the negative of the image. Also, the properties of dilation are just the opposite of the properties of erosion.

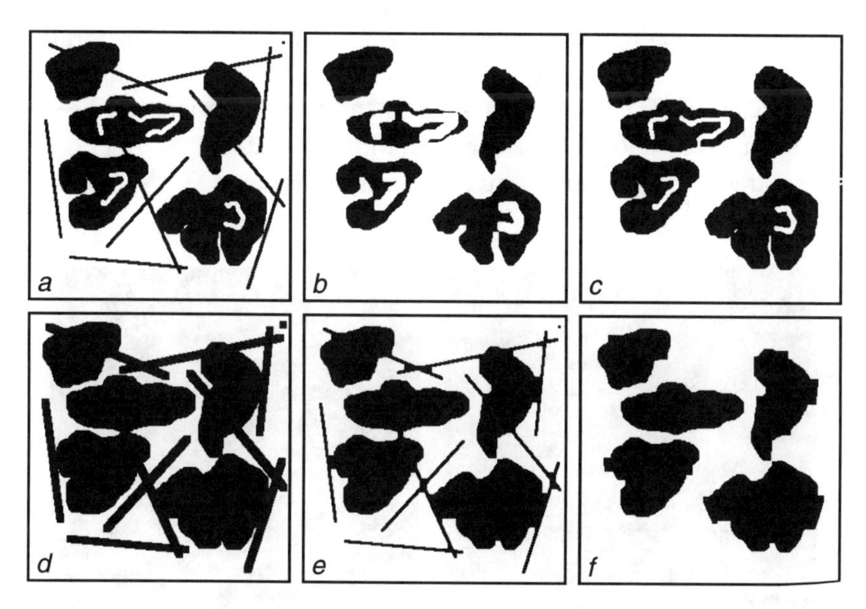

Fig. 2.17. Illustration of the properties of transformations based on erosion and dilation. Initial image (a), the same image after erosion (b), opening (c), dilation (d), closing (e) and closing followed by opening (f).

The above-mentioned operations can produce a compound procedure. Erosion followed by dilation produces *opening*, while dilation followed by erosion gives *closing*. Both transformations preserve the original size of the figures analyzed. Opening can open some holes lying close to the boundary, while closing can close small holes within the image. The basic properties of the whole family are shown graphically in Fig. 2.17.

Erosion, dilation, opening and closing are the simplest morphological transformations and are implemented in nearly all image analysis packages. These transformations, in spite of their simplicity, are important due to their two basic attributes:

- this family of operations is very suitable for introductory cleaning of the image from artifacts and some types of noise

- good understanding of these simple transformations enables understanding of more complex morphological operations.

Applying the above transformations to gray images will give similar effects and we will demonstrate them in practical applications.

Transformations from the erosion-dilation group can be easily used for more complex operations. We will present one example, called the *top-hat* transformation. It detects local minima *(black top-hat*) or local maxima (*white top-hat*). If we observe the results of this operation on the profile line, the application of a top-hat transformation is similar to cutting off the peaks, producing a shape similar to a Mexican hat. The top-hat is a sequence of three operations:

- closing (black top-hat) or opening (white top-hat)
- subtraction of the initial and transformed images
- binarization.

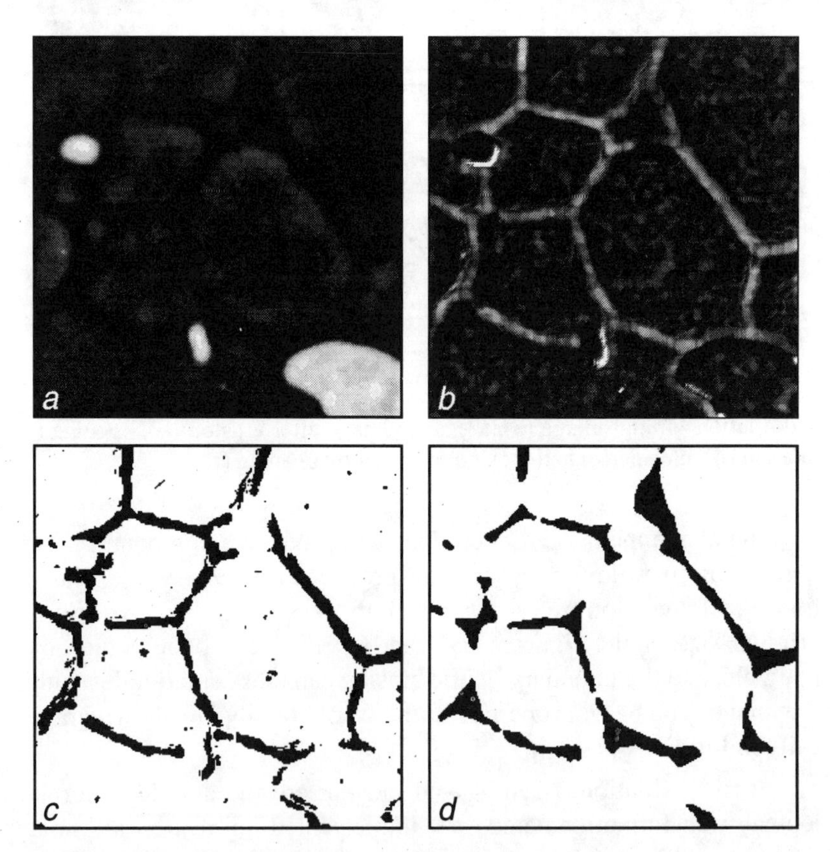

Fig. 2.18. Application of a top-hat transformation for edge detection. Initial image (a), difference between closed and initial images (b), edges detected by top-hat (c), edges detected by binarization (d).

An exemplary top-hat transformation is presented in Fig. 2.18. In this example one can compare grain boundaries detected by a top-hat transformation and simple thresholding. It is clearly visible that top-hat gives relatively narrow and continuous edges, whereas binarization offers a set of isolated islands. In the case of highly non-homogeneous illumination the advantage of a top-hat transformation is even more evident.

Any computer software has very little ability to analyze the shape of any figure. In fact, describing shape is difficult for humans as well. Usually, many objects change shape simultaneously with changes in size. For example, small city cars exhibit silhouettes entirely different from large highway cruisers. Nevertheless, analyzing the shape of a collection of pixels in a computer image is hopeless. It is much easier to describe a set of lines - one can compute and interpret such parameters as: number of branches, loops, end points or crossings, local curvature, etc. Thus, to enable such an analysis we provide a kind of simplified caricature of the figure called a *skeleton.*

One can intuitively understand the idea of skeleton from Fig. 2.19. However, a more formal definition can be helpful for further analysis. We can treat a skeleton as a set of central points of all the disks inscribed into the figure (i.e., disks totally included in the initial figure touching its boundary at two or more points). There is also another approach possible to define the skeleton, as a central line, i.e., a line whose points are equally distant from two closest points of the figure edges.

Skeletons are very convenient for shape analysis, as they preserve a lot of properties of the initial figure:

- skeleton is homotopic with the original figure. This means it preserves the connectivity and number of holes
- skeleton is related to the particle size. It is always fully included within the initial figure and never exceeds this figure. Skeletons of large particles occupy large areas and have long branches, skeletons of small particles occupy small areas and have short branches
- the more complicated and curved the boundary line, the more branches can be noted within the skeleton.

The definitions of skeleton presented above describe an ideal, mathematical skeleton. In the digital case we can obtain only a more- or-less exact approximation (see Fig. 2.20). In some cases (the triangle in Fig. 2.20) the digital skeleton is practically identical with the mathematical model, but in other cases (the circle in Fig. 2.20) it is only a very rough approximation - instead of a single point we get at least four branches touching the circle edge.

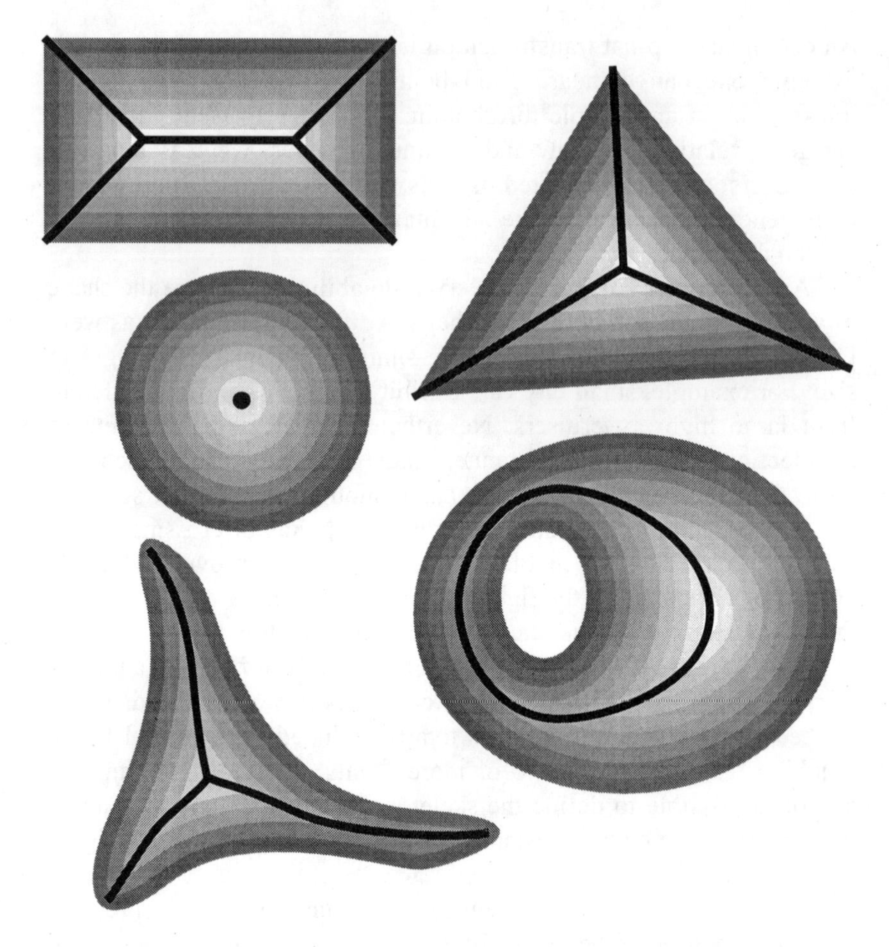

Fig. 2.19. Some examples of geometrical figures and their skeletons (thick black lines). Different gray levels in the figures are applied for better visualization of the distance from the edges.

For the digital case we can build a skeleton as a series of HMTs with a rotating, specially designed structuring element, for example:

X	0	X
X	1	X
1	1	1

In order to get correct results, the above element should be rotated 90° after every step of the transformation. Obviously, other structuring elements or their combinations are allowed. All the structuring elements leading to thin (one pixel-thick) skeletons can be applied. The transformation process is repeated until idempotence, which means to the moment when the next step does not alter the image.

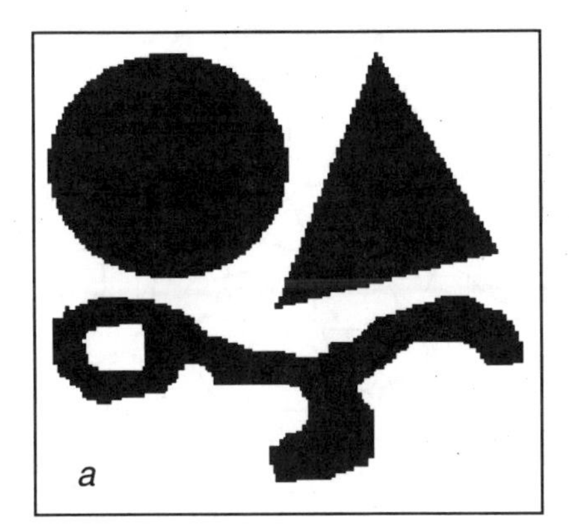

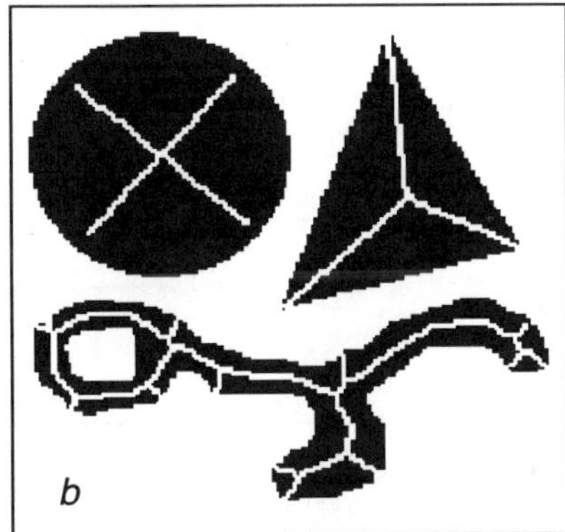

Fig. 2.20. Example figures (a) and their digital skeletons (b).

There exists another widely used tool, tightly coupled with skeletonization, called SKIZ (SKeleton by Influence Zones). Let us imagine a set of points in the image. The influence zone of a given point is a set of all the points lying closer to this point than to any other one. SKIZ can be recognized as a skeleton of the negative of the initial image with subsequent removal of all unnecessary branches.

SKIZ is applied in the analysis of grained structures and is a basic tool for simulation of microstructures. Applying SKIZ to the set of random points (Fig. 2.21a) leads to simulation of the microstructure according to Voronoi partition (Fig. 2.21b). This algorithm gives very impressive results but the process is static - all the points are defined at the very beginning. It leads to grain size distribution far from that met in real materials.

During solidification we observe two simultaneous processes: growth of the existing grains and creation of new seeds, leading to an increase in the grain number and changes in the grain size distribution. Proper tuning of the rates of both processes allows one to obtain size distributions similar to real cases. This variant of simulation is known as the Johnson-Mehl process. An example can be seen in Fig. 2.21d. Continuous addition of new points during simulation is decisive for the dynamic character of this process, making it suitable for research in the field of solidification, diffusion, grain growth, recrystallization, etc.

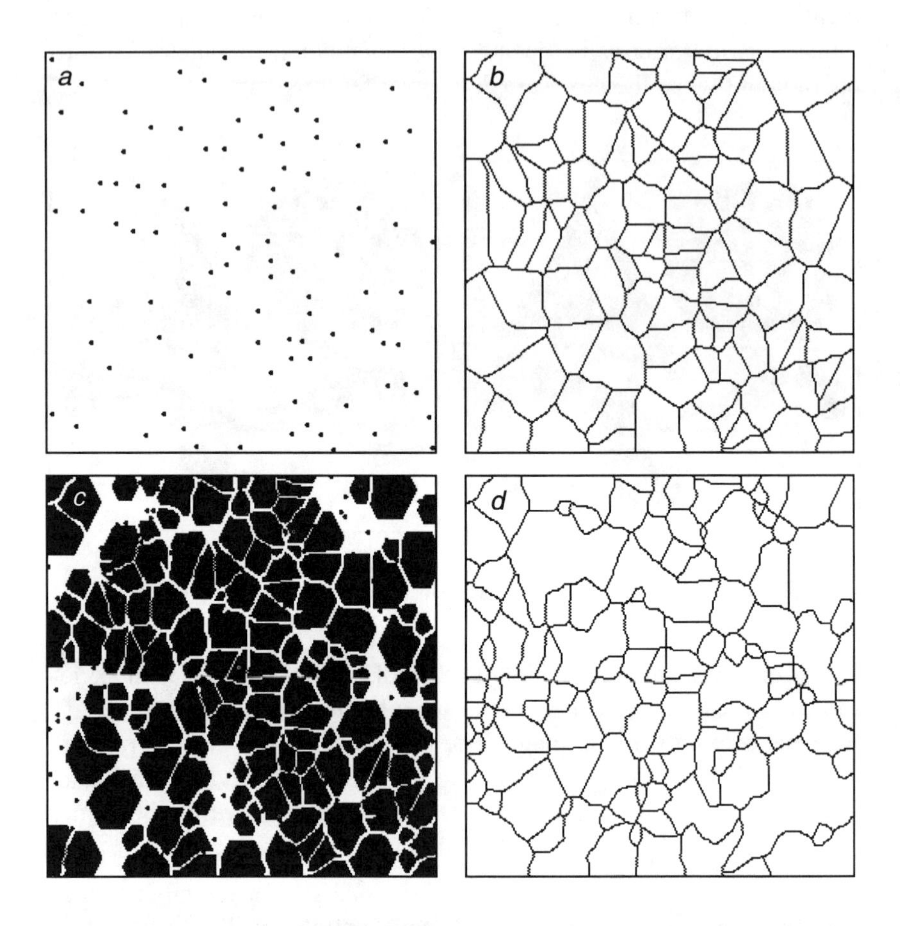

Fig. 2.21. Application of SKIZ for simulation of the single phase structure. Initial set of random points (a), the Voronoi partition simulated by SKIZ (b), grain growth process locked before reaching idempotence and additional random points playing the role of grain seeds (c) and the final simulated microstructure, closer to reality (d).

The last morphological transformation described here will be *watershed* detection. This transformation cannot be applied directly in binary images; it requires an image with at least a few gray levels. The gray-scale image can be interpreted as a map indicating local height as an appropriate gray level. In such a map, the brighter pixel the higher the point in the imaginary terrain. Watershed detection is just finding on the map continuous lines connecting local maxima and is equal to detection of physical watersheds. So, thanks to the map analogy, it is very easy to understand intuitively the idea of watershed detection (see Fig. 2.22).

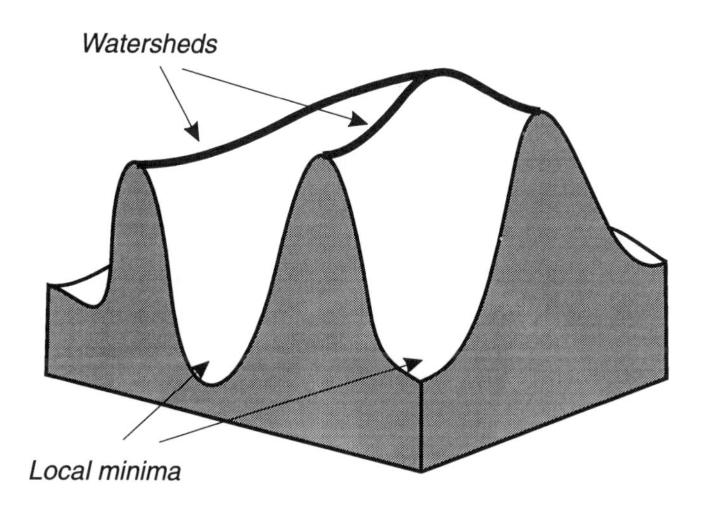

Fig. 2.22. The idea of watershed detection.

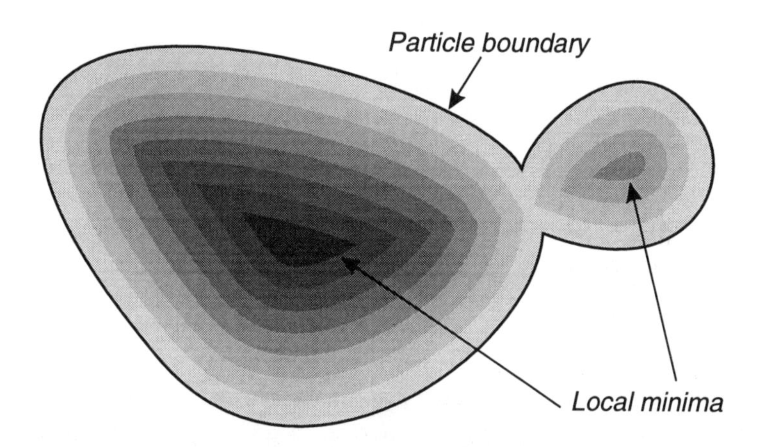

Fig. 2.23. Distance image built on the basis of binary image of a concave particle. More precisely, it is a negative of the distance image, as the particle center is its darkest point.

Binary images can be used for watershed detection after transformation into *distance* images. The distance image (sometimes called *distance function*) is an image with gray levels of individual pixels proportional to their distance from the particle boundary. It can be easily obtained by summation of the subsequent erosions. In the case of concave particles (see Fig. 2.23), we often apply the negative of the distance image. In such circumstances watershed detection allows division of the concave object into two separate, convex particles (see Fig. 2.24).

The above analysis explains two main fields of application of watershed detection:

- separation of particles glued together
- restoration of lost or non-etched grain boundary lines.

In more advanced algorithms watershed detetection can be applied as an efficient tool for edge detection or extraction of isolated particles with blurred edges. Thus, watershed detection is a very strong tool for image analysis, giving results impossible to obtain with other methods.

2.6 Fourier transformation

It can be proven that any continuous function *f(x)* can be effectively expressed as a summation of trigonometric functions of increased frequency. The Fourier transform[80] of the function *f(x)* defines the amount of each frequency in the spectrum and is written as

$$F(u)=\int_{-\infty}^{+\infty} f(x)e^{-2\pi(iux)}dx \qquad (2.3)$$

where:

$i=\sqrt{-1}$

The so-called Euler formula enables interpretation of the Fourier transformation as a sum of trigonometric functions:

$$e^{-2\pi(iux)}=\cos 2\pi(ux)-i\sin 2\pi(ux) \qquad (2.4)$$

It is very important for practical application of the Fourier transformation that it is reversible, which means that having the transformation we can perform the inverse operation and recover the initial image:

$$f(x)=\int_{-\infty}^{+\infty} F(u)e^{2\pi(iux)}dx \qquad (2.5)$$

The function *f(x)* can represent spatially varying gray levels in the image and takes real values. The transform function *F(u)* is complex and has two parts: real *R* and imaginary *I*.

$$F(u)=R(u)+I(u) \qquad (2.6)$$

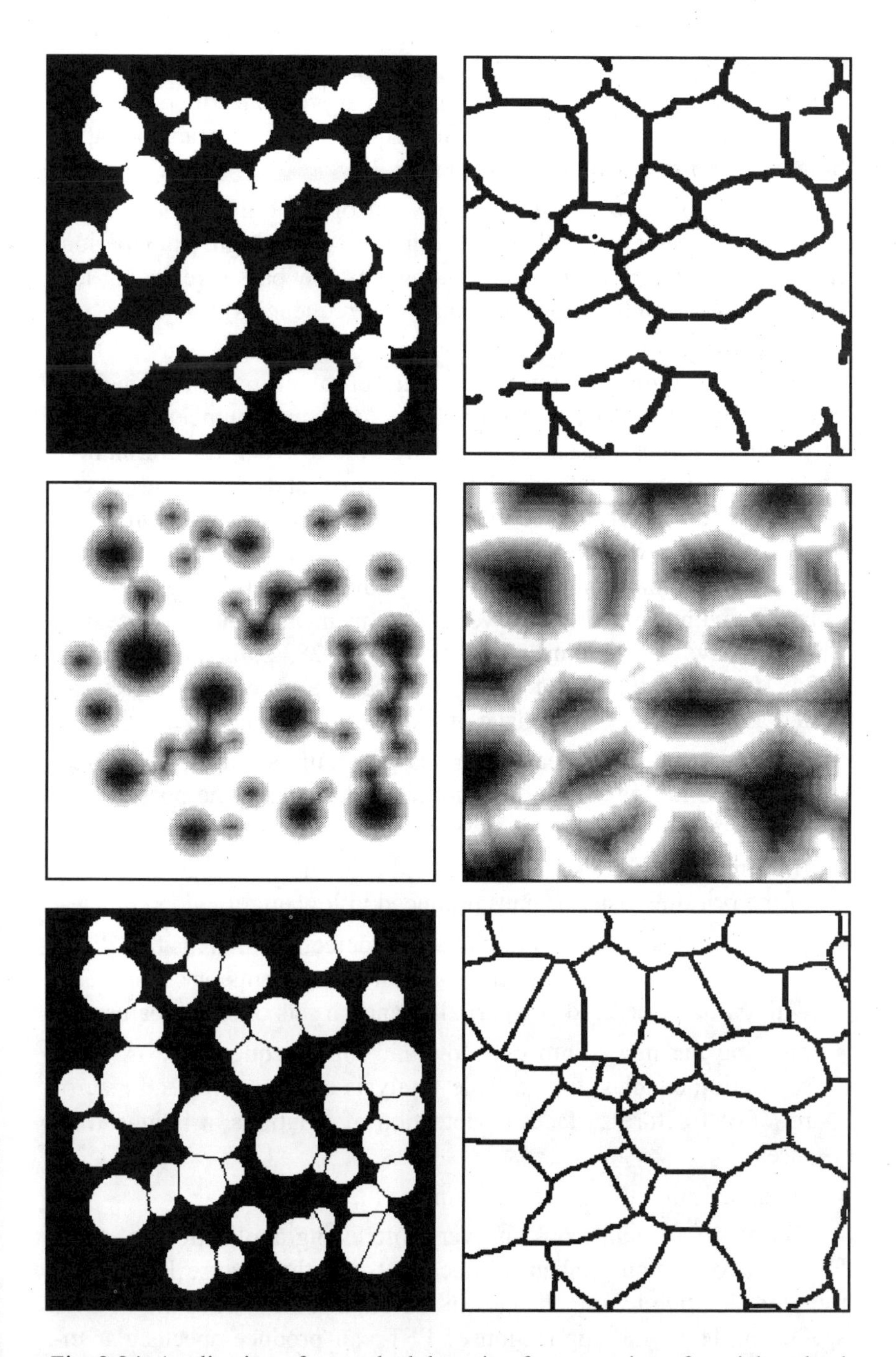

Fig. 2.24. Application of watershed detection for separation of particles glued together (left column) and lost grain boundary restoration (right column). From top to bottom: initial image, distance image and final result.

Fourier transformation consists of a series of trigonometric functions (see Eqn. 2.4), having their own amplitudes and frequencies. One can modify (in a direct way or using appropriate filters) the amplitude for a given frequency. Subsequent inverse transformation builds the transformed image (see Fig. 2.25).

Usually the modified part (mainly cut off) is only a small part of the Fourier spectrum, with low (long period) or high (shot period) frequency. Appropriate filters are known as low pass (preserving low frequencies) or high pass (preserving high frequencies) filters.

Obviously, in the case of digital images, Eqns. (2.3 to 2.6) are modified into a discrete form. The main problem in the practical application of the Fourier transformation is the huge number of necessary computations that take a very long time to reach transformation. Fortunately, there exists a relatively simple and sufficiently exact approximation which can be efficiently computed. It is known as the *fast Fourier transform* (FFT) and just this form of transformation is used in image analysis packages. FFT, however, can be applied only to images with a size of the power of 2. This means that for successful FFT we have to use images of size 128x128, 256x256, 512x512, 1024x1024, etc. Images of other sizes can be processed if they arc pasted as a part of a blank image of a size suitable for FFT.

The nature of FFT means that it is especially sensitive to any features that exhibit any form of periodicity. So, among the possible applications of FFT one should stress the following:

- filtering out any noise of a periodical character, for example, traces of the printing raster or lines produced in low quality video images
- texture analysis in fine structures, i.e., detection of areas that differ only in orientation of extremely small components or constituents with partially ordered and partially random sets of small features
- dividing the image into two, low and high frequency parts. Such separation enables, for example, analysis of the micro- and macro-relief of fracture surfaces or detection of striations in fatigue fractures
- advanced smoothing or sharpening of the image.

So, there are numerous and very interesting applications of FFT. However, one should taken into account that the inverse FFT is extremely sensitive to any changes in the amplitudes within the Fourier spectrum. Thus, the application of FFT can produce spectacular effects, not available with the help of other image analysis tools but, on the other hand, it requires a lot of time-consuming work to prepare an efficient application.

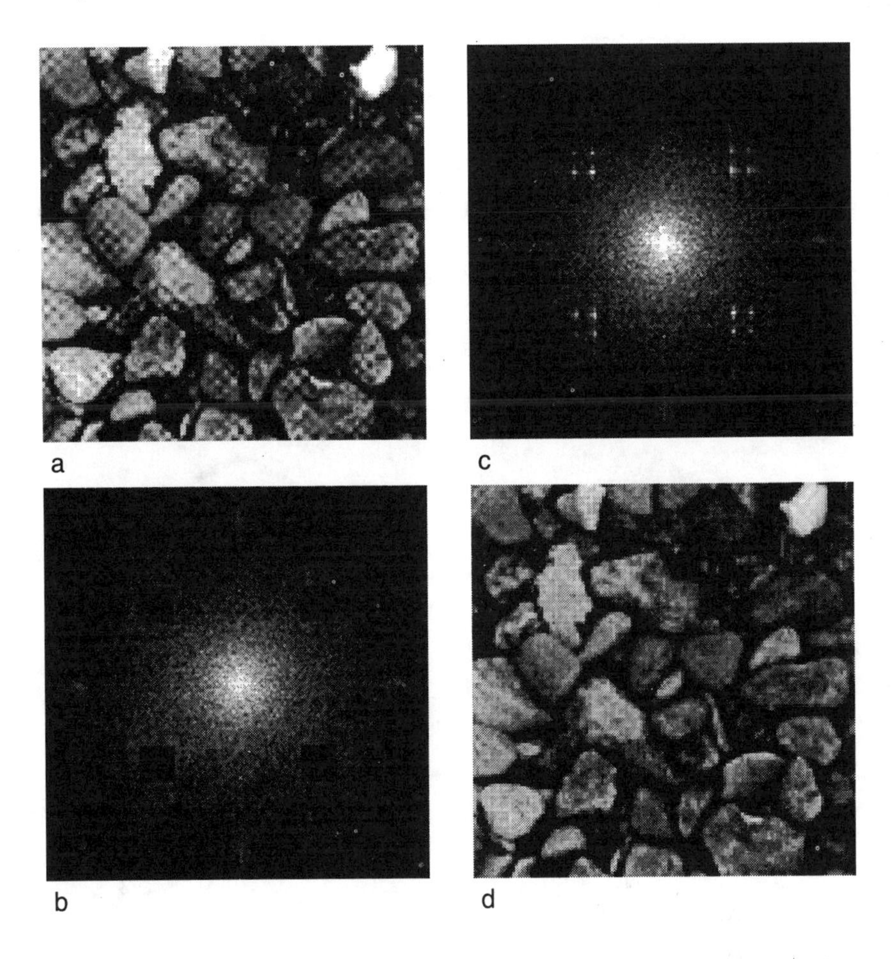

Fig. 2.25. Example of the Fourier transformation. Initial noisy image (a), Fourier spectrum (b), the same spectrum with peaks produced by the noise suppressed (c) and filtered image after inverse FFT (d).

An example of a Fourier transformation, applied to remove some noise of periodical character that was introduced by the printer raster, is shown in Fig. 2.25. The initial, noisy image of some gravel particles is shown in Fig. 2.25a, whereas its Fourier spectrum is presented in Fig. 2.25b. The noise creates four groups of peaks in the Fourier spectrum, visible as white points. These points can be effectively suppressed without alteration of the remaining part of the spectrum (Fig. 2.25c). Inverse FFT (Fig. 2.25d) gives an image with the noise well-filtered. Such an outcome cannot be effectively obtained with other tools.

2.7 Edge detection

Edge is a physical or imaginary line marking the outer limit or boundary of a surface. The human vision system is especially sensitive to recognition of edges. Thanks to this phenomenon, accompanied by the ability to analyze visual data in 3-D, we can safely walk, drive, eat, etc. Similarly, in computerized image analysis edge detection is one of the basic tools for extracting features being detected from the original image. However, in contrast to the human, computerized edge detection is not straightforward.[15,18,21,34,80,84,114]

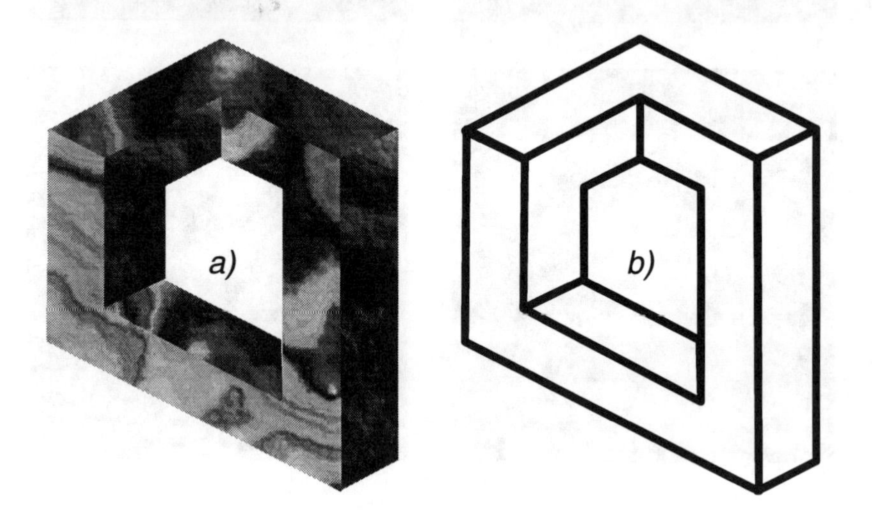

Fig. 2.26. A figure (a) and its edges drawn manually - it can be done by everyone of us (b).

Let us have a look at the figure drawn in Fig. 2.26a. Due to some patterns in this surface, the edges of this figure are not very clear. In spite of it, nearly everyone could immediately draw the sketch representing the edges, as shown in Fig. 2.26b. In a later part of this chapter we will discuss some tools which can be helpful to perform similar edge detection.

There exists, however, the next interesting property of the sketch shown in Fig. 2.26. The system of lines in Fig. 2.26b suggests some three-dimensional character of the figure, but simultaneously there is something wrong in the perspective. Pictures drawn with this property are called impossible figures. This graphical effect has been well known for centuries. Unfortunately, it is practically impossible to detect this property using contemporary image analysis tools.

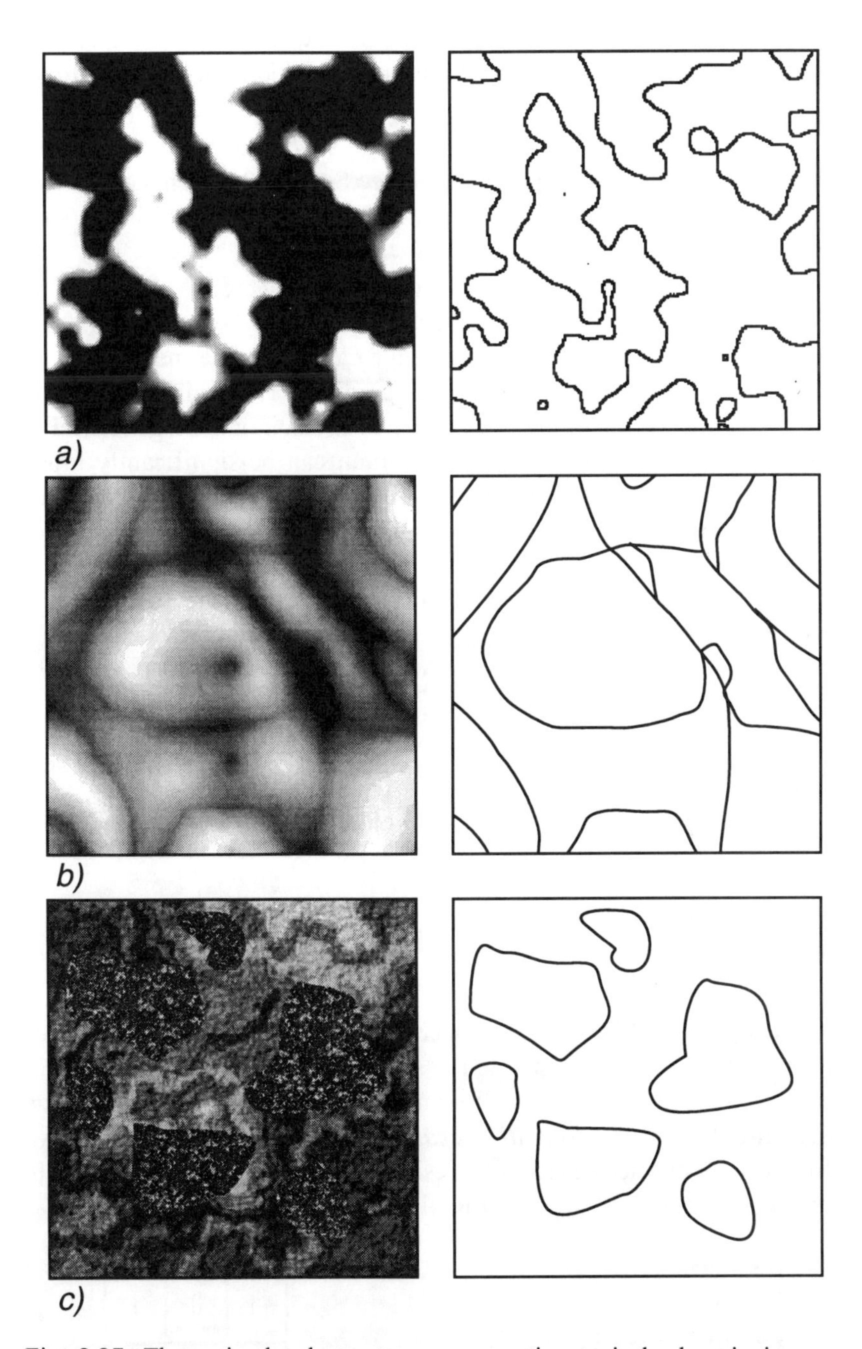

Fig. 2.27. Three simulated structures representing typical edges in images (left) and detected edges (right): the boundaries between regions of different gray levels (a), edges as local minima in gray levels (b) and boundaries as lines separating regions of different textures (c).

Fig. 2.27 presents three of the most frequent types of structures met in practice, in which one should detect edges. Now, we will discuss these cases in more detail.

The simplest case is shown in Fig. 2.27a and represents edges possible to detect using binarization (see Section 2.4). This case takes place in the analysis of well-contrasted particles embedded in a matrix or in the analysis of some multiphase materials.

The most common case for edges to be detected is illustrated in Fig. 2.27b. In this case edges can be recognized as local minima or maxima of gray level. This type of edge can be also defined as a place with the highest value of gray level gradient. The main difficulty in the detection of edges of this type lies in their inhomogeneity. For example, the value of the local maximum can be significantly lower than the value of a local minimum situated elsewhere within the same image. Thus, there exists no universal tool for detection of such edges and among the existing procedures everyone can be the best, depending on the case. We will briefly discuss the following functions for edge detection:

- top-hat filter
- Prewitt operator
- Roberts operator
- Sobel operator
- Laplacian filter
- zero crossing operator
- watershed detection.

The third type of edges, based on different textures in the neighboring regions (shown in Fig. 2.27c) is described in Section 5.1.

The *top-hat filter* was described in Section 2.5 and an example of its application in edge detection is shown in Fig. 2.18 (the same sample image, as in Fig. 2.18, will be used for presentation of other edge detection tools). The *Prewitt operator* is a typical filter, or rather two filters, with matrices especially designed to amplify local gradients in the gray level in the Y and X directions, respectively:

1	1	1
0	0	0
-1	-1	-1

and

-1	0	1
-1	0	1
-1	0	1

An example of Prewitt filtering is shown in Fig. 2.28b. It is noteworthy that in contrast to the top-hat filter, the Prewitt operator tends to produce dual edges. This results from the fact that the boundary line in the initial image (Fig. 2.28a) is relatively thick.

A similar, but somewhat different approach was proposed by Roberts. The *Roberts operator* is, in fact, a series of four filters, detecting very short (single pixel-long) edges with four angles of orientation: 0, 90, 45 and 135 degrees.

0	0	0
0	-1	1
0	0	0

0	0	0
0	-1	0
0	1	0

0	0	0
0	-1	0
0	0	1

0	0	0
0	-1	0
1	0	0

The Roberts operator works in four directions, but it takes into account only a single point, whereas in the Prewitt's solution three neighboring pixels were analyzed (in two perpendicular directions). As a consequence, the Roberts operator produces edges similar to Prewitt edges, but finer and more delicate (see Fig. 2.28c). Note that the Roberts operator produces numerous, very short edge segments, clearly visible in Fig. 2.28c.

Another edge detection operator, very similar to the Prewitt filter was introduced by *Sobel*. The only difference between these two approaches is that Sobel replaces one pair (1, -1) by a pair (2, -2). It results in a little coarser edge detection (see Fig. 2.28d):

1	2	1
0	0	0
-1	-2	-1

-1	0	1
-2	0	2
-1	0	1

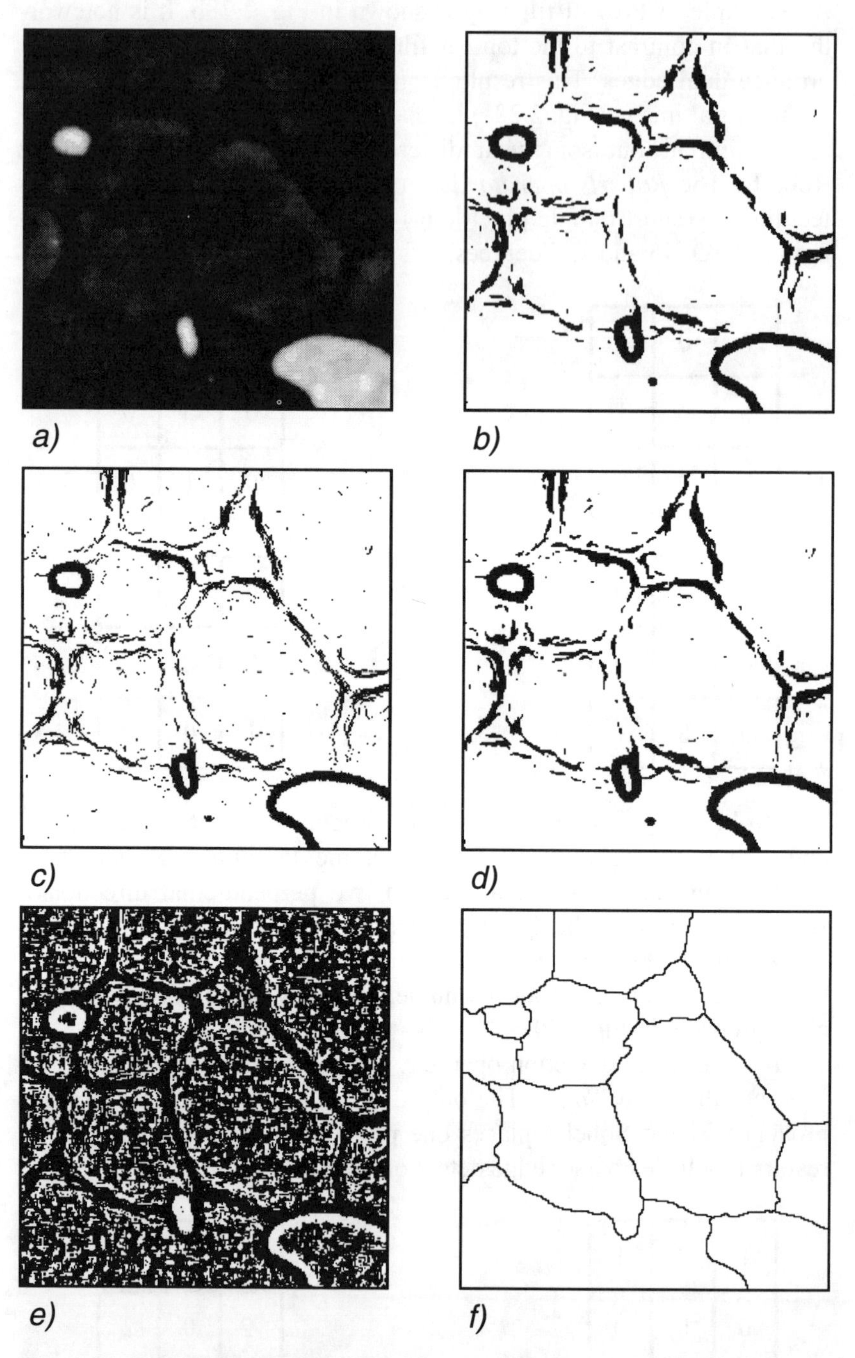

Fig. 2.28. Examples of edge detection operators: initial image (a), Prewitt filter (b), Roberts filter (c), Sobel filter (d), Laplacian (e) and watershed-based edge detection (f).

The next approach is known as *Laplacian*. It is a filter similar to sharpening filters, being an approximation of a Laplace operator. In the simplest case it uses the following matrix:

-1	0	-1
0	4	0
-1	0	-1

Application of this filter gives zeros if there is no gray level gradient and values different from zero for other cases. The absolute value after operation is largest for pixels with the highest local gradient in gray levels. An example of this filter is shown in Fig. 2.28e. This gives a good starting point for further analysis - it is enough to convert the regions with dense white spots into a continuous white surface. Unfortunately, practical realization of this simple postulate is very difficult and often impossible.

All the filters discussed above detect very well the edges of very bright, well contrasted regions in Fig. 2.28a. Obviously, this could also be easily obtained using the simplest threshold, as in this case we have edges of the first type, shown in Fig. 2.27a.

It is also clear in Fig. 2.28b-e that edge detection filters produce broken, non-continuous edges. This seems to be their largest disadvantage, as we usually need just continuous edges for further analysis. In order to get such boundary lines, one can use a *zero crossing filter*. This filter uses Laplacian as a starting point and puts an edge at every point where Laplacian changes its sign (or crosses zero, which is exactly the same). Such an approach produces high quality, continuous edges, but under the condition that the Laplacian image also detects the edges clearly. Unfortunately, this is rarely the case.

The last approach is based on *watershed detection* (see Section 2.5 and Fig. 2.24). Application of a watershed operation requires a lot of supplementary transformations and long-lasting, laborious fine tuning. It is especially difficult to obtain a universal procedure capable of automatically and correctly detecting edges in a whole series of images, not only in a single case. How to do this will be shown in more detail in Section 4.1. Here, we can state that this approach, in spite of its complexity, can give really satisfactory results (see Fig. 2.28f).

2.8 Combining two images

In the procedures discussed so far in this chapter we have started with a single image and the procedure transforms it into another image. Now, we will show some simple but useful operations that produce a new image, starting from two, not one, initial images. The operators discussed here can be divided into two groups:

- logical functions
- arithmetic operations.

In both cases we have a basic restriction that the initial images have to be of the same dimensions. In other words, they should have an identical number of pixels along the vertical and horizontal directions.

We will not analyze any properties of *logical transformations.* Instead, we will demonstrate them using simple examples. Let us take two binary images, denoted in Fig. 2.29 as A and B, respectively. In Fig. 2.29 one can note six different logical operations:

- *negative* of image A (NOT A); the only logical transformation that has only a single argument (image). Other operations use both initial images - A and B
- *logical sum* of images A and B (A OR B)
- *logical intersection* of images A and B (A AND B)
- *exclusive or* (A XOR B) - a function that returns differences between initial images
- negative of the previous operation, i.e., *not xor* (A NXOR B). This function returns identical parts of both initial images and
- *logical difference* (A - B)

Logical functions are frequently used in image analysis when we need to create a new image from those already existing. A typical example is shown in Fig. 2.30. Our task is to separate three particles glued together (Fig. 2.30A). After appropriate treatment of this image, for example, with application of watershed detection, we get image B, representing dividing lines (Fig. 2.30B). In order to produce the final set of separated particles we should use logical functions. The solution chosen here is A AND (NOT B). From the sketches illustrating basic operations in Fig. 2.29 it is quite easy to check the correctness of this solution and try to find another, equivalent solution.

One should be aware of applying logical transformations to gray level images. It is possible, but the whole operation is performed bit by bit, so the results are strange and unpredictable. Thus, it is safer to use logical operations *only* for the treatment of binary images.

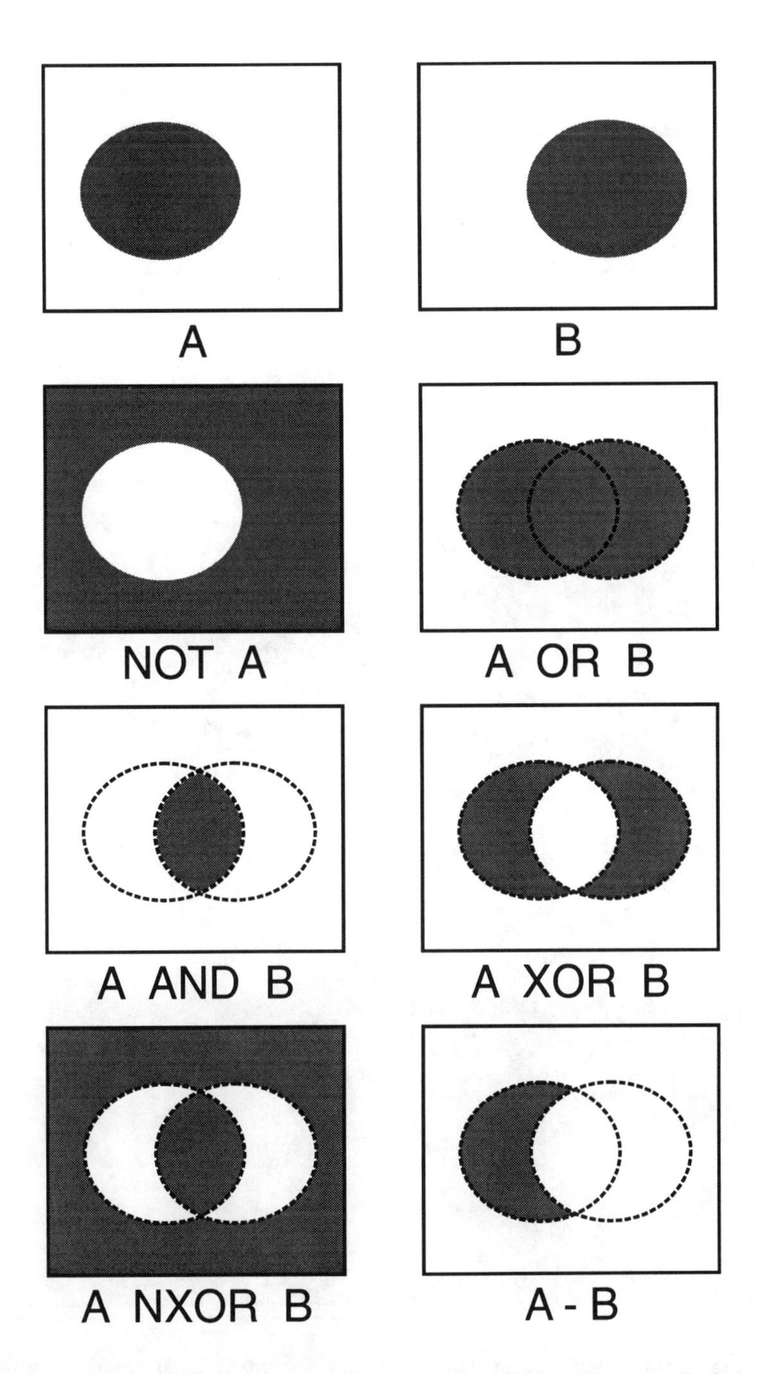

Fig. 2.29. Schematic illustration of various logical operations applied to two initial binary images: A and B.

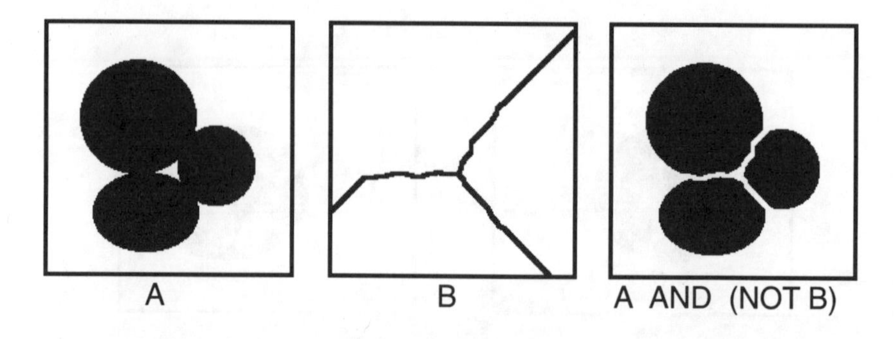

Fig. 2.30. An example of the application of logical operations for dividing the glued together particles.

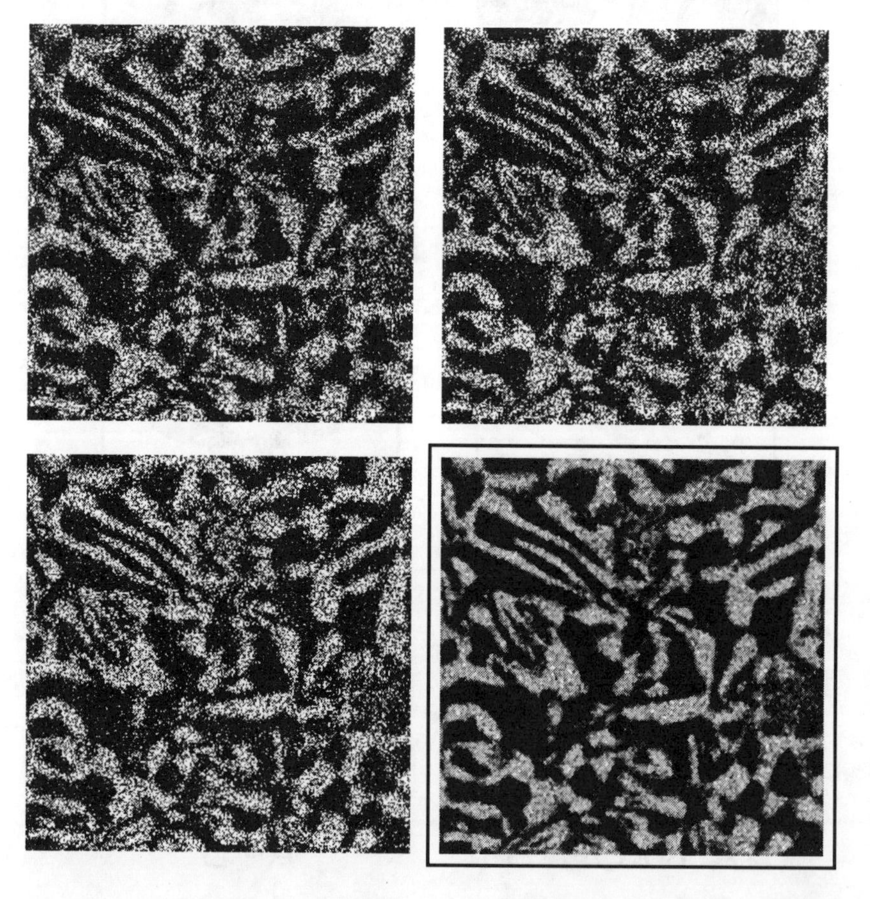

Fig. 2.31. Simple but tricky application of arithmetic operations for noise reduction. If we have a lot of poor quality images with random noise we can try to add them and get a much better result (see the lower right-hand corner).

Arithmetic operations are much easier to understand than logical ones. If we add two pictures, the resulting image will have gray values as simple arithmetic sums of appropriate pixels in the initial images. Similarly, we can perform multiplying, subtraction or detection of minima and maxima between two images. The only problem is caused by division - the initial values are usually integer numbers, whereas the final result is usually a floating point. This requires appropriate formatting of data and more memory.

Arithmetic operations seem to be quite easy to understand. In spite of their simplicity, arithmetic operations can be very useful. One tricky example is shown in Fig. 2.31. In some cases we have such dark images that the noise produced by CCD elements results in poor quality, noisy images. Fortunately, this noise has, in general, random character. Thanks to this last property we can try to take numerous snapshots of the same object and add the images. Noise should be well compensated. Obviously, we have to normalize the final sum in order to prevent getting enormous values of the resulting gray levels.

2.9 *Mouse etching*

If we are able to indicate manually the features to be detected, it can be done automatically, as well. The only problem is the time necessary for finding an appropriate solution. In very simple and rare cases it can take some hours. Usually it takes days or weeks. In very complex cases it can take years - an example is the continuous effort to detect cancer cells in medicine.

In some cases it is worthy or necessary to devote a lot of work in order to invent an automatic procedure for detection. Examples of these events could be:

- routine control procedures - their automation can save a lot of time and, consequently, money. Moreover, this can increase the accuracy and repeatability of the measurements
- high end research and optimization - automatic methods can prevent gross errors as we often see subconsciously what we want, not what really exists
- features that are important but difficult or impossible for clear detection or classification (it is typical in medical applications). A good example would be restoration of lost grain boundaries.

If you have no time, no money and no idea how to solve your problem and only a few images to be analyzed, try to detect the interesting features manually. With today's hardware it can be relatively easy to do with help from a computer mouse. Therefore, this process is sometimes called humorously *mouse etching*. This technique, however, preserves a lot of advantages of fully automated image analysis:

- it enables the use of many tools for image enhancement (sharpening, noise reduction, shade correction, etc.)
- all the measurements can be done automatically even on manually prepared samples.

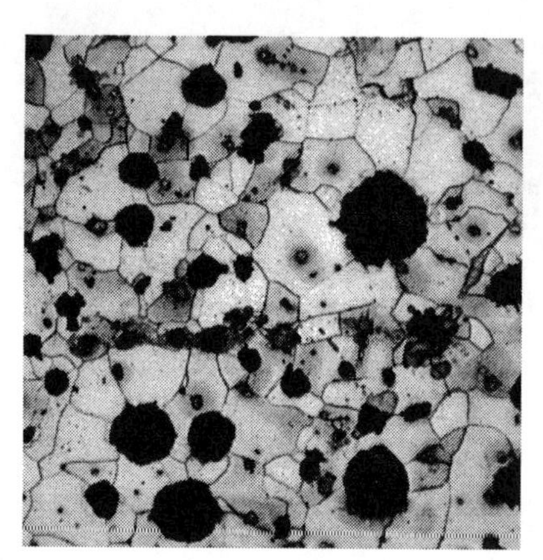
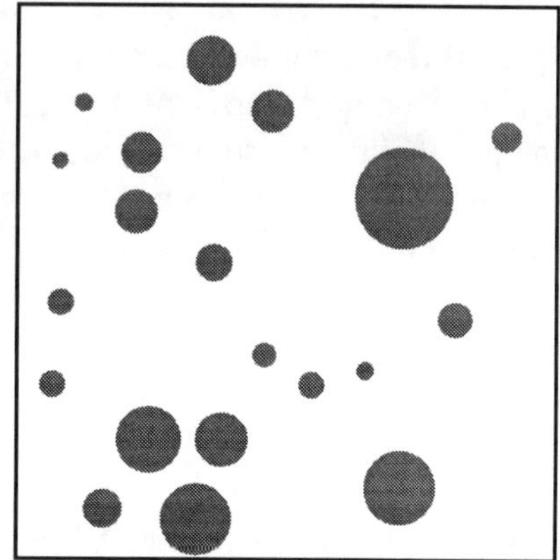

Fig. 2.32. An example of a low quality image of nodular iron, which cannot be used for automatic image analysis due to high contamination (left). Manually detected graphite nodules (right) can be used for automatic size distribution measurements.

This last point is probably the key one. People are always better at *interpretating* images (or cars would be driven without any difficulty by computers), whereas computers are unbeatable in making *measurements*. Thus, mouse etching can sometimes lead to optimal use of the tools offered by computerized image analysis (see Fig. 2.32).

With these remarks we will close this chapter, devoted to the main tools for image treatment. Obviously, these tools were described only in a rough form; for further details the reader can consult the specialized literature. On the other hand, this shortened and often intuitive description should be sufficient for those who have no time or ability to study the details of mathematical theories and algorithms. It should support them in understanding what is really done within the computer memory and encourage them to built their own algorithms.

Chapter three

Image acquisition and its quality

3.1 Specimen preparation

In most cases analysis of materials structure is done with the help of microscopes of various types. Consequently, careful preparation of specimens is necessary as it is decisive for the quality of images obtained. The quality of the initial image is decisive for the results of analysis, affecting both the ability to detect features under analysis and the precision of subsequent measurements.[5,45,83,89,90,113] The value of proper specimen preparation cannot be overestimated.

The majority of materials are opaque and can be observed in optical microscopes only in reflected light. Specimens for this type of observation are usually prepared by mechanical grinding and mechanical or electrolytic polishing. The general rule is that hard materials are relatively easy to polish whereas soft materials require special techniques and care for acceptable results. Some very hard materials, for example, ceramics, are also very difficult for correct specimen preparation. Due to their hardness they are practically non-workable and their high chemical resistance disables electrochemical polishing.

Obviously, there is no place here to discuss methods of specimen preparation - suitable data can be easily found in numerous handbooks on metallography. In this chapter we will discuss only some selected aspects of specimen preparation that are interesting from the point-of-view of image analysis. Thus, we will focus on possible errors that should be avoided if the given specimen is to be used for automatic processing.

Especially difficult for correct preparation are very soft materials, sensitive to smearing or non-homogeneous materials, containing both very soft and very hard constituents. A good example of this type of material is ferritic gray cast iron (Fig. 3.1). Graphite is extremely soft, ferrite also is very plastic and soft whereas phosphorous eutectic, being hard and brittle, tends to produce some relief, even in the early stages of grinding on papers (see white arrows in Fig. 3.1). Such errors cannot be successfully removed during further polishing. Large scratches introduce high plastic deformation, which can affect the results of etching, even if the geometry of the polished section exhibits perfect quality.

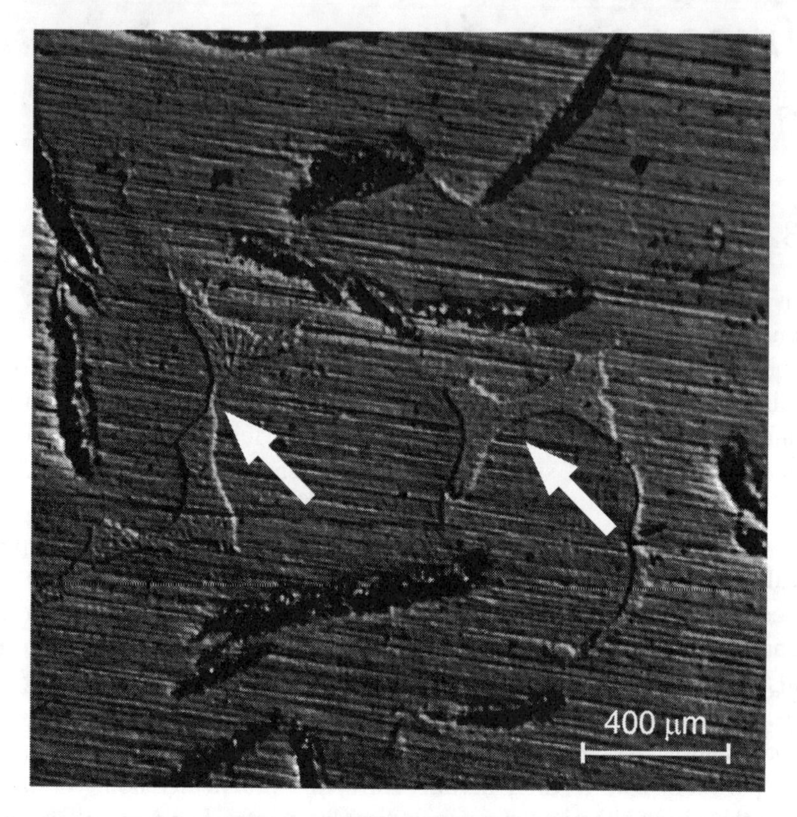

Fig. 3.1. Roughly prepared (grinding on papers) section of ferritic gray iron. The structure is very soft except for the phosphorous eutectic which produces some relief (white arrows).

Relief introduced by hard structural constituents is also visible in Fig. 3.2, this time on a finely polished section. Additionally, black arrows indicate some relief at the particle/matrix interface. This is a very frequent inaccuracy, notably in the case of precipitates softer than the matrix. In order to avoid this type of relief one can try to use the following procedures:[3]

- avoid too delicate grinding
- try to prepare the polished section quickly
- avoid large specimen sections. Good results are easier to obtain on smaller specimens
- the mount for the specimen should be large and not too soft - it helps to keep the polished section flat.

In case of really strong difficulties the best solution is to contact an experienced technician. There are many materials which are really very difficult for correct specimen preparation and only after tens of trials is it possible to find out a suitable polishing technique. Sometimes only the use of awkward and unusual grinding and polishing materials leads to correct results.

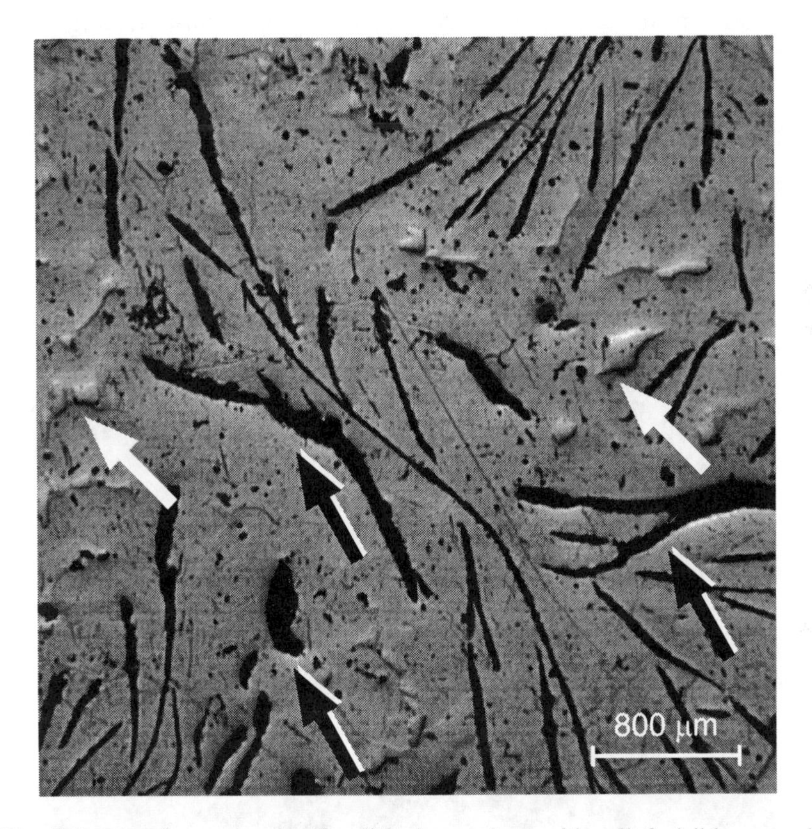

Fig. 3.2. Another example of polished gray iron with relief visible around both soft particles (black arrows) and hard eutectic (white arrows).

Small scale relief is practically always present on mechanically polished specimens, but in the optical microscope, especially in the bright field, it is practically invisible. Fine changes in the specimen height, introduced by relief, can be used for some types of special observation techniques, like interference contrast. Such contrast can provide interesting results but should be always applied with great care.

The next defect, very common for polished specimens, is the presence of scratches (see Fig. 3.3). Scratches can be easily avoided in the case of hard materials, but are very common for soft or multiphase materials with soft particles. Scratches are remains of the introductory grinding or badly performed polishing (dirty polishing disk, abrasive material with too large grains, incorrect pressure during polishing, etc.). If scratches are introduced mainly during polishing, they can be relatively easily avoided. Deep scratches, being the remains of the introductory grinding are either very difficult or impossible to remove. Any attempt to remove them by lengthening the final polishing process leads to extensive relief formation, known also as over-polishing. In some cases, for example, Cu alloys, one can almost completely avoid scratches by applying electrolytic polishing. This does not mean, however, that we can skip careful preparation of the surfaces to be polished.

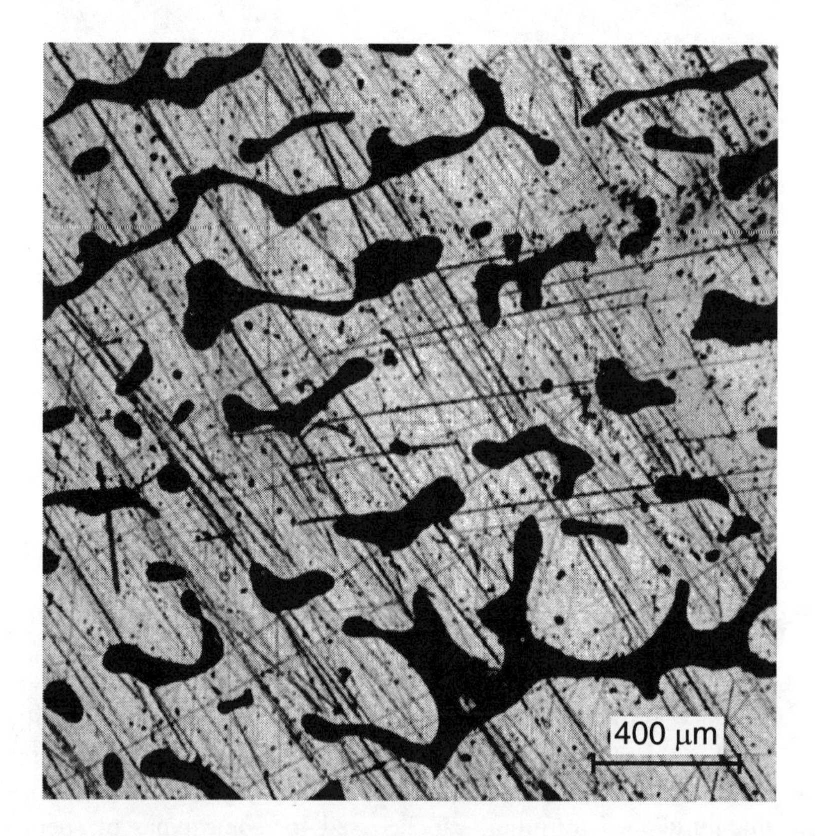

Fig. 3.3. Deep scratches from grinding, observed in the polished and etched section of a soft Cu-12% Al alloy (as-cast condition).

Scratches and relief are not the only possible side effects of mechanical or electrochemical preparation of specimens for microscopic inspection. Freshly polished surfaces are sensitive to any type of chemical reactions, especially oxidation. Corrosive blooms can be formed even during polishing, when water or other solvents used are not pure but contain some traces of aggressive reagents. In the majority of cases, polished specimens are etched in order to reveal grain boundaries or other structural constituents. The specimens have to be thoroughly rinsed before drying to prevent unnecessary etching. Unfortunately, some traces of etching substances tend to be stored in the micropores of the specimen and can produce some stains on the specimen surface (Fig. 3.4).

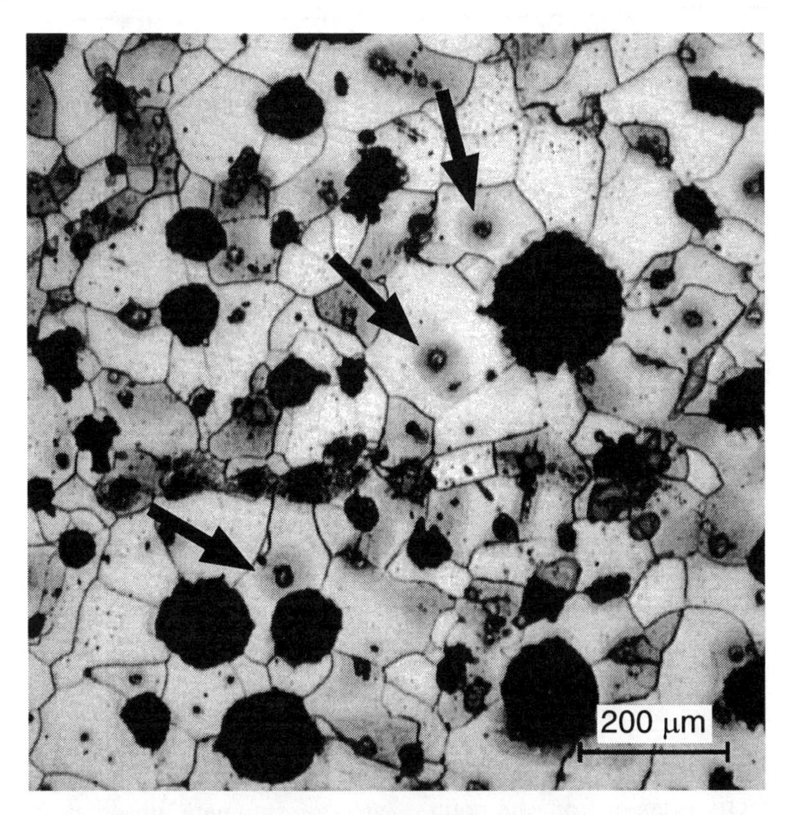

Fig. 3.4. Ferritic nodular iron after polishing and etching. Black arrows indicate evident traces of corrosion visible in the vicinity of non-metallic inclusions. This image was taken only 10 minutes after etching.

Spots, smudges and other traces of corrosion deteriorate the quality of the initial image, often making it useless for further analysis as these traces cannot be effectively removed using image analysis tools. There is a very subtle difference between an optimally etched and destroyed specimen surface. The same etching procedure can give different effects depending on polishing methodology. For example, higher pressure during polishing can increase the velocity of etching, leading to overetching.

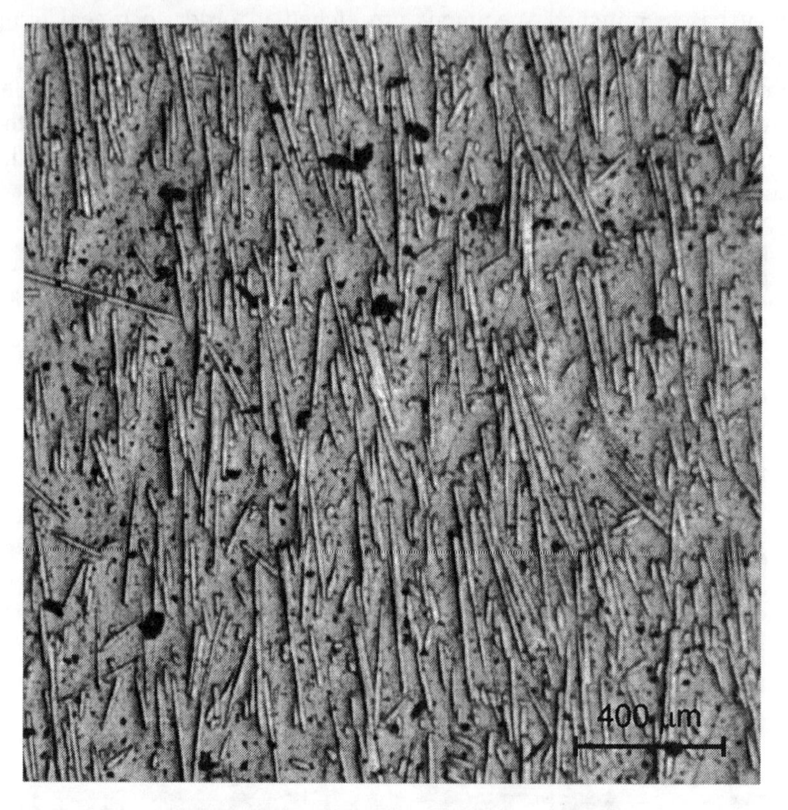

Fig. 3.5. Glass fibers in a thermoplastic resin, un-etched. Note poor contrast between fibers and matrix, making the specimen unsuitable for automatic analysis.

One can also find some materials that are practically impossible to correctly prepare from the point-of-view of automatic image analysis (Fig. 3.5). This can be caused by high chemical resistance, disabling any etching. This is the case in ceramic materials and some composites, as shown in Fig. 3.5. Specific chemical composition can prevent getting proper contrast in optical microscopes as well as in SEM. In such cases one can try to use manual methods for extraction of features for further analysis. Fortunately, current microscopies have such a wide spectrum of visualization methods that correct results can be obtained for the majority of materials and their structures.[27,109]

To summarize these remarks on specimen preparation, one can put forward the following conclusions:

- image analysis can extract some information from the image but cannot replace careful specimen preparation
- any image devoted to automatic image analysis should be of the highest possible quality. Only high quality initial images enable automatic analysis of long series of specimens with sufficient precision
- for good reproducibility of specimens (and, consequently, results of the further analysis), automatic grinding and polishing are advised whenever possible
- many errors that are easily interpreted and corrected by a human observer cannot be successfully processed by a computer
- special care is necessary during examination of both soft and multiphase materials.

3.2 *Image acquisition*

Any image, seen in the microscope or taken by any type of camera or other device, has to be transferred into computer memory. This process is called image acquisition and can be performed in several ways:[21,114]

- one of the oldest is data input from a *digitizing tablet*. This device is relatively slow and inaccurate as the data input is done manually and depends on the precision of the operator. On the other hand, a digitizer can be a good solution in the case of images with very poor contrast or high noise level and for the input of linear features, like fracture profiles, etc.
- *a scanner* is a good and relatively cheap solution. It is the best solution if the images for analysis are available in the form of photographs or hard copies of some other type. Scanners work relatively slowly, so they are not a good solution for massive data input
- *a CCD video camera* is currently one of the most frequently used sources of data for computer image analysis. It is a smart device, offering good sensitivity and speed, being standard equipment on many optical microscopes. Some problems arise if we want to analyze color (an RGB camera is advised since the VHS system gives poor resolution of colors) or high resolution images because appropriate cameras and adapters (*frame grabbers*) are very expensive. Moreover, every frame grabber requires special software to

be coupled to the computer. So, one should be very careful when choosing hardware and software. Special care should be focused on the compatibility of the system components. Video recorders (VCR) can be used interchangeably as an equivalent of any CCD camera since the output data format of VCRs is identical to that of CCD cameras

- contemporary *electron microscopes* are equipped with modules for digital storage of images. Therefore, there is no problem with transmission of the images from the electron microscope into the image analysis programs. Moreover, many leading companies assembling electron microscopes include image analysis packages as standard in their products.

So, there are numerous ways of data input. However, the majority of problems are common for all these methods and therefore can be discussed independently from the source of data. For example, secondary or backscattered electron modes in electron scanning microscopy can be compared with different contrast methods in light microscopy. The analogy of bright and dark field in optical and electron microscopy is even deeper.

An interesting problem of the proper choice of magnification is illustrated in Fig. 3.6. We see in this picture the microstructure of concrete and the image quality is acceptable for automatic processing. In spite of this we cannot analyze the size distribution or shape of the gravel particles, as only a small region of the gravel particles is visible. There are two solutions in such a situation: we can either decrease the magnification to get a larger field of observation or take a series of pictures touching each other and subsequently reconstruct the whole structure. On the other hand, an even larger magnification could be necessary for inspection of fine features on the inter-particle interface, for example.

Proper choice of magnification is always an important problem and requires careful analysis. There are no rigid rules in this matter, however, some guidelines may be extracted:[21,44,45,80]

- be careful that the magnification used does not exceed the resolving power of the microscope. In such a case you can get so-called *empty magnification*, giving an image with the smallest details larger than a single pixel
- if you want to test inhomogeneity, keep in mind that at high magnification every structure is inhomogeneous. The linear size of the image for this purpose should be a few times larger than the expected radius of region defined as homogeneous. If you apply any

tools of stereology for analysis of inhomogeneity, you should carefully follow the requirements of the method described

- if you want to analyze some particles, the suitable magnification should lie somewhere between the following two limits: the smallest particles should contain at least approximately 10 pixels in order to enable any reliable shape or size measurements (lower limit) and the largest particles should be somewhat smaller than the linear size of the image (upper limit). Obviously, some particles will always be cut by the image boundary, but this phenomenon can be effectively corrected.

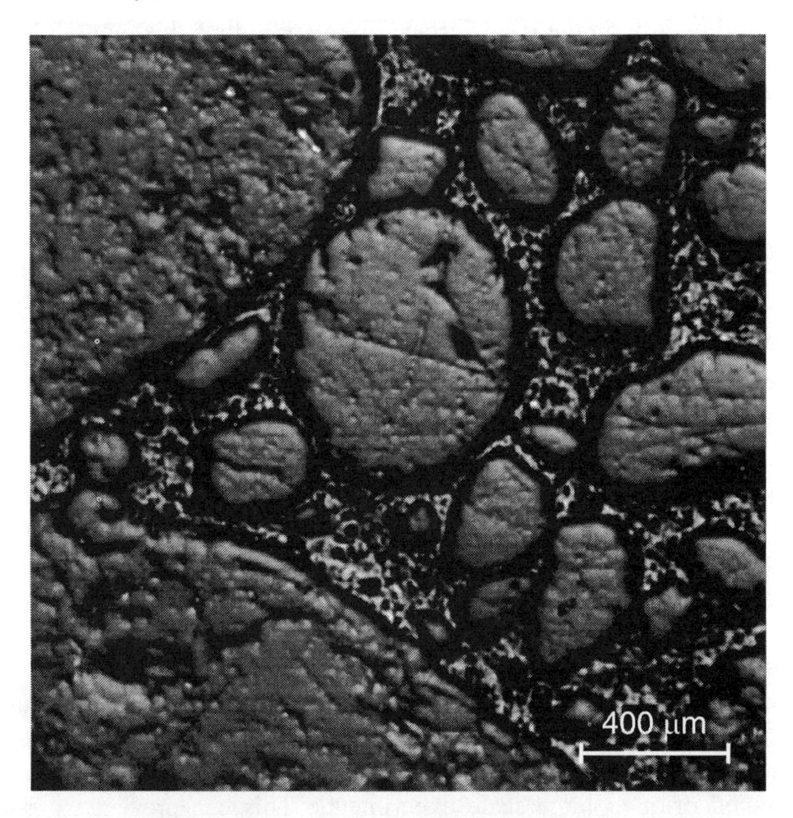

Fig. 3.6. Microstructure of concrete. Magnification applied is too high - only a portion of the gravel particles is visible, making size distribution analysis impossible. An example of a poor image in spite of acceptable preparation quality.

The key problem in image acquisition is grabbing images that contain just the necessary information. In the case of optical microscopes, this is done mainly by proper etching or illumination; whereas in scanning electron microscopy (SEM), it is by the proper choice of the signal used for image formation. There are two main signals used: backscatterd electrons (BE) and secondary electrons (SE). Fine tuning of these signals allows one to get the desired information.[27,35,39,65,78, 94]

The image taken in BE has some contrast based principally on local changes in chemical composition, with some additional information from the local surface topography. This mode gives relatively little information on the local topography of the specimen.

The contrast in SE images is formed in a different way and is similar to that observed in optical microscopy. The local gray levels depend mainly on the local surface orientation, thus this mode is suitable for topographic examination, for example, in porous materials or in fractography.

Comparison between the above described modes is illustrated in Fig. 3.7, on an Al_2O_3-Ni_3Al composite, observed in SE (top) and BE (bottom). If there is a need to detect the individual phases, the SE image is useless for this purpose, whereas the constituent phases can be easily detected in the BE image using a simple binarization procedure.

The SE image is very sensitive to changes in the acceleration voltage (see Fig. 3.8). Correction of the acceleration voltage can alter the resulting image almost completely, including local inversion in the contrast (see central part of the images in Fig. 3.8). Low voltage makes the SE image (Fig. 3.8 bottom) similar to the BE image (Fig. 3.9). Thus, manipulation of the imaging modes, acceleration voltage and specimen preparation (covering by evaporation of thin layers of gold or other chemical treatment) produce countless variants of the same image. Depending on the character of the structure under analysis various types of images can be of the highest value. It is quite common to use various modes at different stages of image treatment. Therefore the technique of image acquisition should be carefully planned prior to large scale experiments. This can prevent us from unnecessary repetition of the experiments.

Electron microscopy also offers other signals, illustrating, for example, surface density of various elements, specimen conductance or some diffraction patterns. Every type of such images contains some information on the material and its microstructure. Our knowledge, experience and imagination or intuition should prompt us on how to use this information for automatic quantification of the analyzed specimens. Some examples will be presented in later chapters.

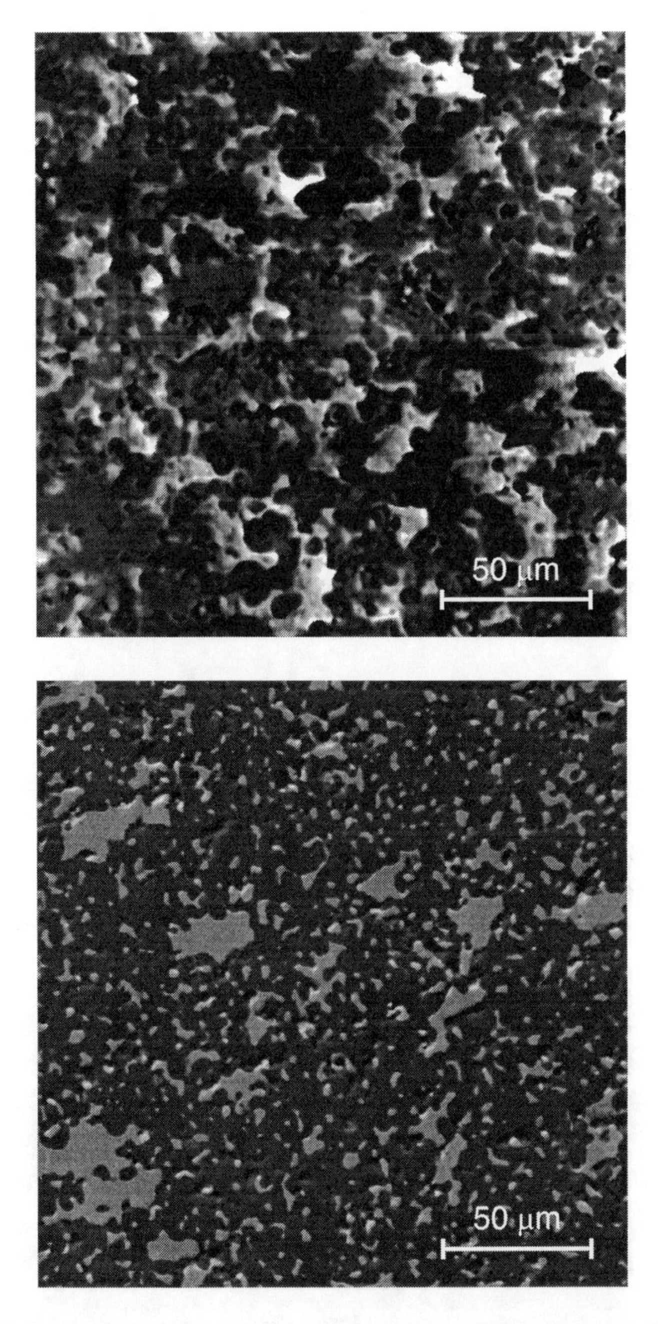

Fig. 3.7. A composite Al_2O_3-Ni_3Al, observed in secondary electrons (top) and backscattered electrons (bottom). Note that only the latter image is suitable for detection of the phases (dark Al_2O_3 and light Ni_3Al).

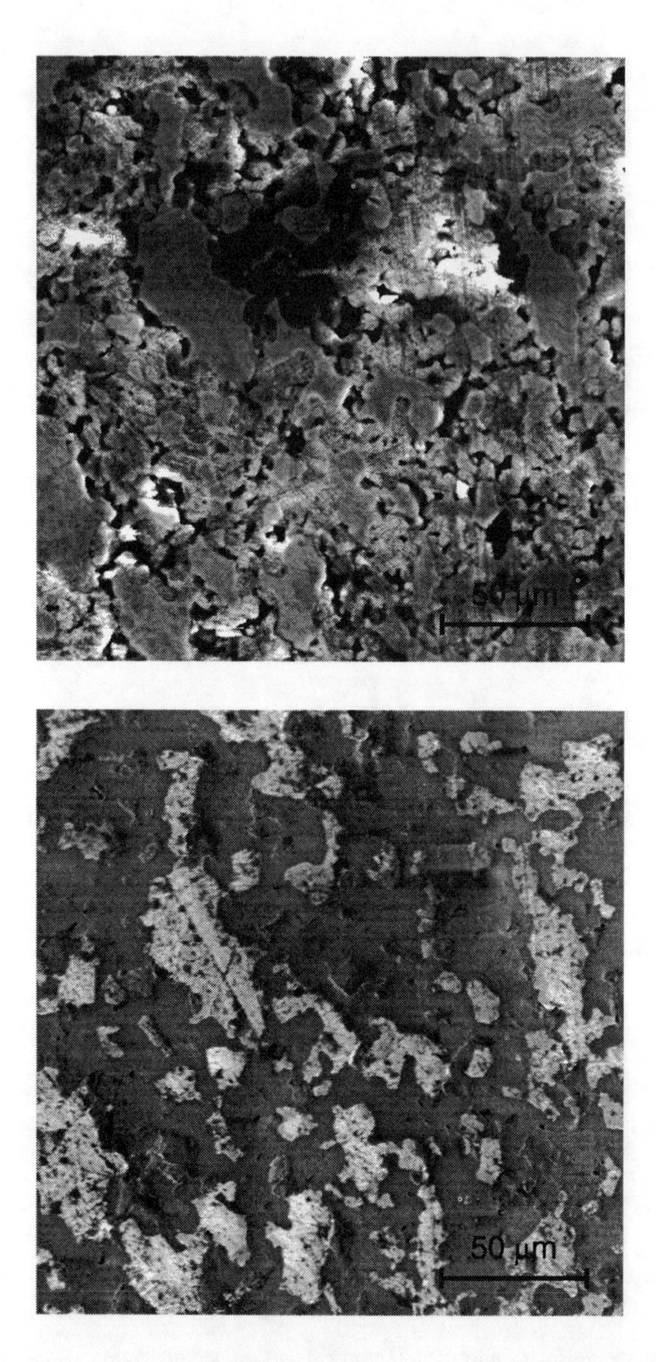

Fig. 3.8. The same Al_2O_3-Ni_3Al composite, as in Fig. 3.7, observed in secondary electrons at different acceleration voltages: 15 kV(top) and 1 kV (bottom). Note that the lower image is similar to the BE image (see Fig. 3.9).

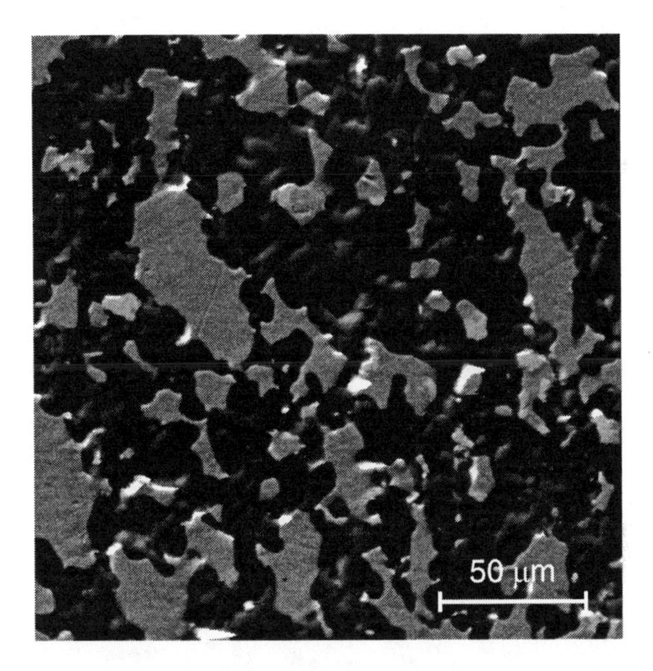

Fig. 3.9. Same region as in Fig. 3.8 but observed in backscattered electrons.

In electron microscopy, especially in the case of scanning microscopes, we have an extremely large depth of focus. This allows examination of rough surfaces (fractures, arrangements of fibers, fabric, etc.), including stereoscopic examination. Unfortunately, there is only limited ability to restore the three-dimensional structure of rough surfaces, as the brightness of the given point tends to vary with the angle of tilt. Thus, finding the adjacent points in the pair of images can be extremely difficult. In the case of analyzing the three-dimensional nature of some structures, much better results can be obtained when using confocal microscopes. They give almost perfect information on the local topography of the specimen.

3.3 Shade correction

Some images, primarily from optical microscopes, exhibit irregular illumination, called *shade*. Some regions (see Fig. 3.10a) are brighter and some others are darker than the mean value for the whole image. This phenomenon is a consequence of inaccuracy in the optical system and, first of all, an inhomogeneous light source. Precise tuning of high quality microscopes can minimize this effect, but in many cases it significantly disturbs the analysis. The main problem caused by the presence of shade is that it can extensively affect the results of binarization, especially of phases with gray levels close to the background of the image. Thus, much effort is expended in order to correct this distortion.[45,80,95,109]

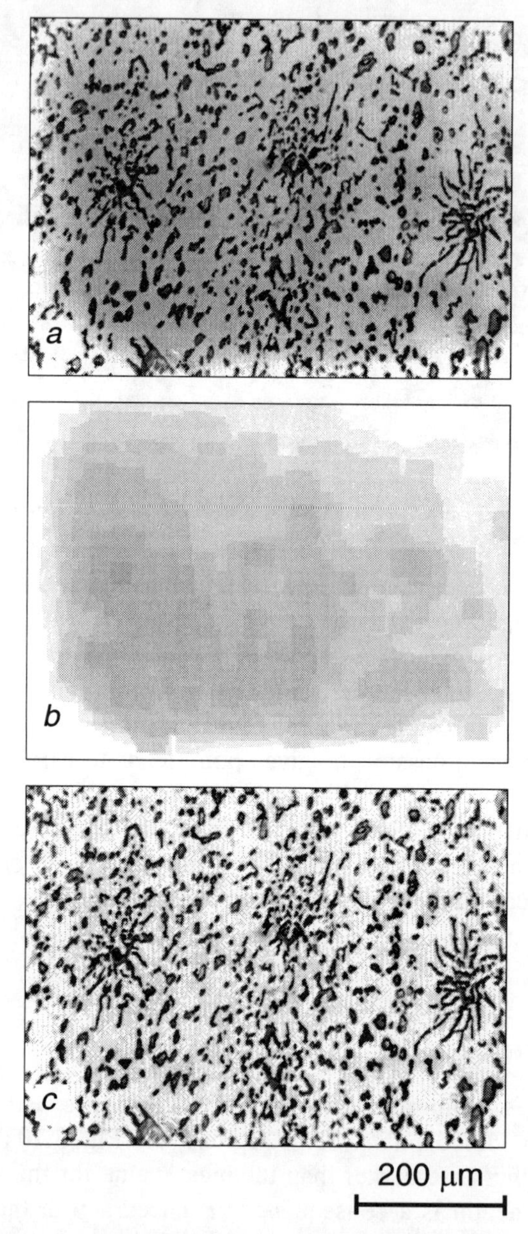

Fig. 3.10. Shade correction. Initial image (a), extracted shade (b) and corrected image (c).

If we compare the initial (Fig. 3.10a) and the final, corrected image (Fig. 3.10c), we come to the conclusion that shade correction can be effectively completed. This is true, but shade correction requires some care to avoid typical errors leading to false results (unfortunately, looking correct at the first glance). The simple and efficient correction procedure is thoroughly explained and described below.

The first, most important and difficult step is extraction of the shade, which can be defined as an image illustrating the inhomogeneity of the light source. In some image analysis systems it is possible to calibrate the system - the shade is detected from the image of a perfectly flat mirror and appropriate hardware performs suitable correction. Images obtained from such an instrument are free from shade. Unfortunately, we get images with more or less clear shade quite frequently and we face the need to extract it from the image.

In most cases, at least in light microscopy, shade is recognized as uneven brightness of the brightest part of the image. It is visible especially clearly in images with a low content of dark elements. Grain boundaries of single-phase materials and pores or inclusions in unetched polished sections are typical examples of structures whose shade is easy to detect.

How to extract the shade? Assuming that the shadowed background should be white (we can always use the negative of the image if its background is dark or use complementary transformations), we can use one of the following solutions:

- close the image; the size of the closing should be large enough to remove all the dark parts - pores, grain boundaries, etc. This method cannot be applied to images with large dark areas
- perform an FFT transformation, apply a low pass filter to the Fourier spectrum (shade consist of low frequencies) and afterward inverse the FFT to get the image of the shade. This method should be used with care, especially in the case of the presence of large dark particles. The result of this form of shade extraction should always be controlled by the operator in order to avoid abnormal results
- (difficult and sophisticated) you can mark the brightest points, choosing, for example, local maxima, and build the image using interpolation methods. Interpolation methods construct an artificial image in such a way that any profile is a smooth line and the values of pixels located at the positions of their prototypes - here local maxima - have the same gray levels as these prototypes. In some image analysis or computer graphics packages you can find appro-

priate procedures. If there is lack of them, try to use other solutions because building the effective interpolation tool on your own may be very difficult and laborious

- in the case of unusual shades, try to invent your own solution. For example, you may see an image in which the left side is much brighter than the right side and there is a continuous change in the background level between these sides. In such a case you can perform a linear erosion with a size equal to the image height, using a vertical (90°) structuring element. As a result, you will get a collection of vertical lines with gray levels equal to the darkest point along the line. This will be a high quality shade extraction.

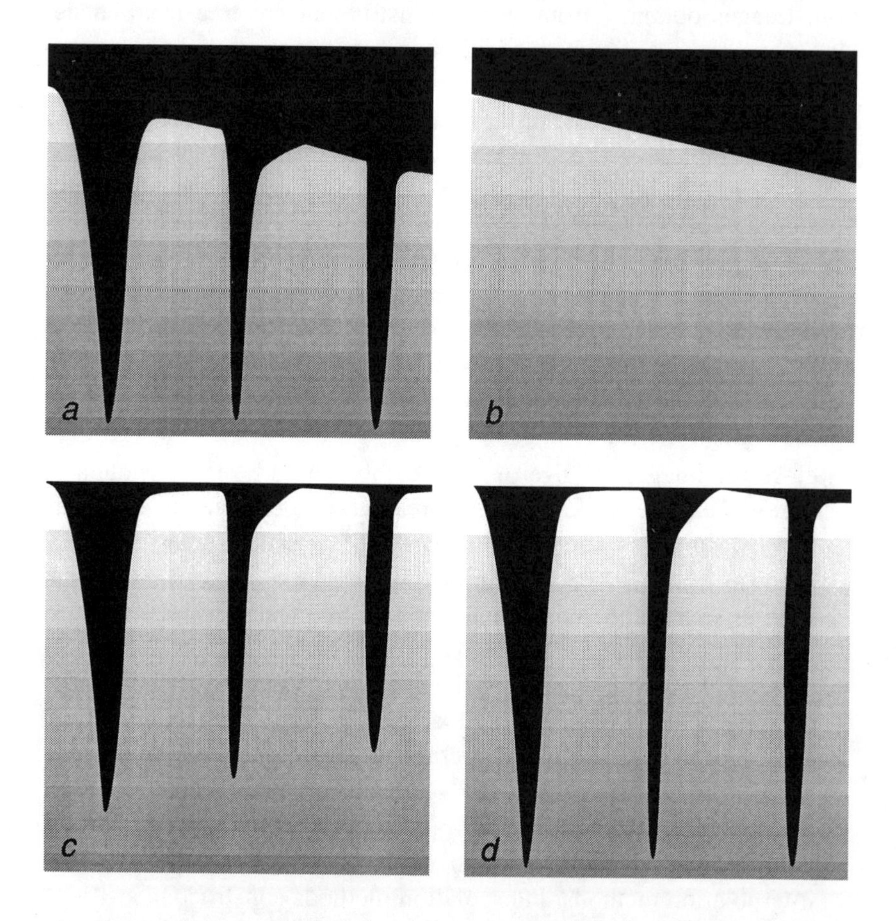

Fig. 3.11. Schematic illustration of the shade correction procedure, Initial profile (a), profile of the shade (b), improper shade correction based on subtraction of the images (c) and proper shade correction based on division of the images (d). Detailed description in the text.

If we have already extracted the shade, it is possible to complete the shade correction. The simplest procedure used is based on image subtraction and addition.[28] You take the following steps:

- create the *temporary* image, subtracting the shade from the purely white image. The result will be complementary to the shade; in other words, adding shade and the temporary image will lead to an image with all pixels white
- create the *corrected* image by summation of the initial and temporary images.

The procedure is very simple and leads to results that at first glance seem to be correct. Unfortunately, closer analysis (Fig. 3.11c) shows that the results are not correct. Adding the temporary image causes all the originally black pixels to be brighter, as they obtain the values of the temporary image. This may lead to false results of binarization, especially if one needs to detect some dark features (see Fig. 3.12).

To avoid the systematic error described above, a new procedure is recommended. In this procedure the same shade is used, but the pixel values change proportionally to their values. This algorithm leads to correct results and can be described in the following way:

- create the *temporary* image, dividing the purely white image by the shade image. The result will have floating point values (ensure before division that the result will have floating point values) showing how many times white is brighter than the gray level of the given pixel in the shade image
- create the *corrected* image by multiplication of the initial and temporary images.

The above described procedure is very efficient; however, it requires a rather more advanced image analysis package, enabling the use of floating point values in the image. There are also some programs with built-in procedures for shade removal, usually based on filtering out the lowest frequencies from the Fourier spectrum. In the case of the presence of shade removing procedures you can use them, but it is advised to check the result carefully, for example, by comparison of the profiles before and after the operation. If the result seems to be wrong, it is always possible to use the procedures described above.

Finally, it should be noted that in the case of weak shading the procedure based on subtraction and summation of images also gives acceptable results. However, it is always a good habit to check the result using the profile (see Chapter 2).

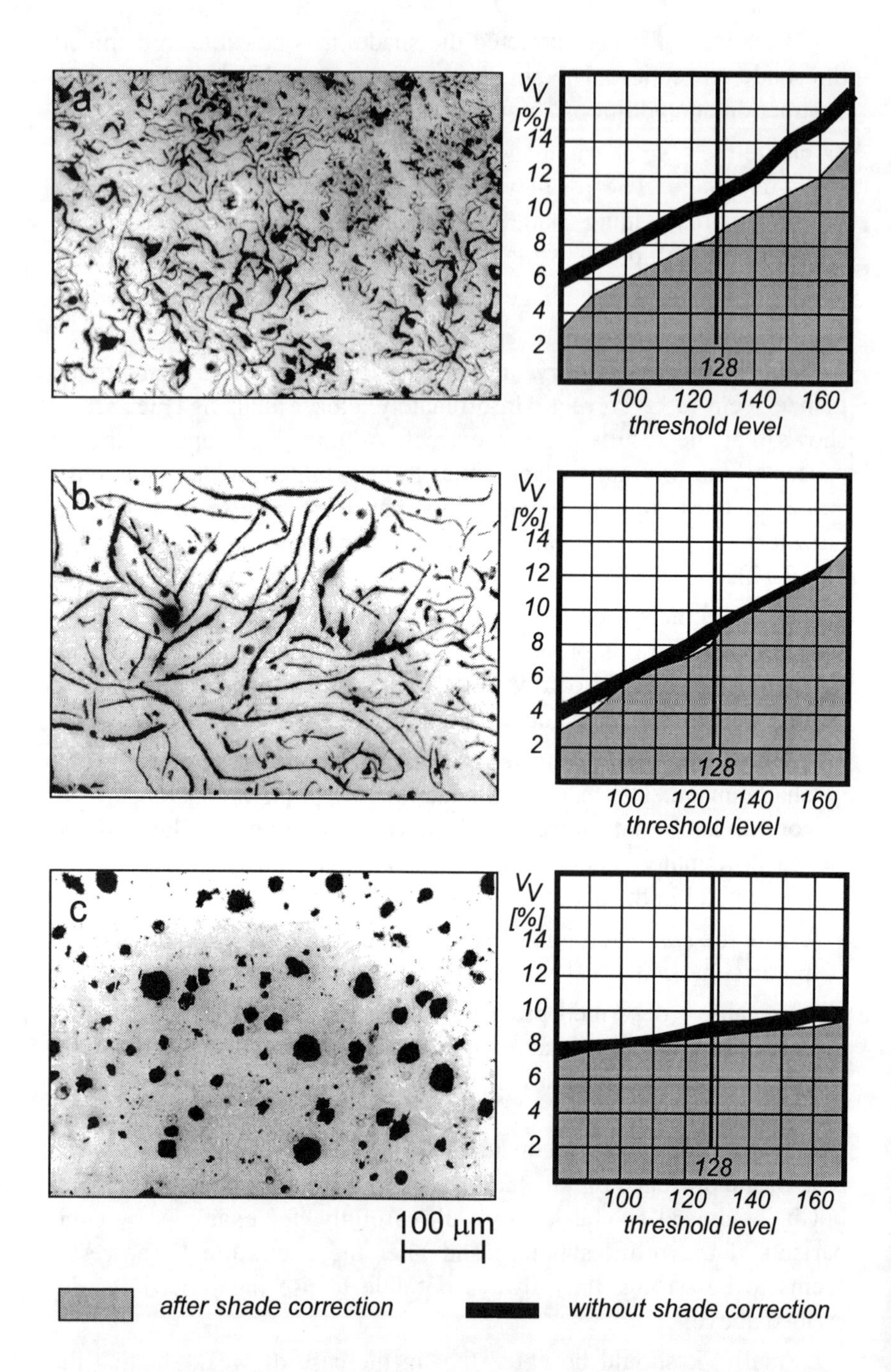

Fig. 3.12. Effect of the binarization threshold and shade correction on the volume fraction of graphite precipitates in: malleable (a), gray (b) and nodular (c) cast iron. The microstructures are shown without shade correction.

Analysis of the results presented in Fig. 3.12 leads to the following conclusions:

- the effect of the shade is particularly strong in the case of fine precipitates (Fig. 3.12a). In this case a relatively large area is occupied by boundary pixels, with gray levels between the values typical for graphite (dark) and matrix (bright), respectively. In such a case the threshold level has to be chosen with special care - refer to Section 2.4 for possible remedies
- the effect of threshold level depends on the shape rather than on the size of the particles (compare the plot in Fig. 3.12c with the plots in Figs. 3.12a and 3.12b)
- the gray level after shade correction was close to 0 for graphite and close to 255 for the matrix, respectively. So, the value of 128, marked in the plots in Fig. 3.12 by a vertical line, is equal to the arithmetic mean of these two gray levels and should be a safe (high probability) choice for the correct threshold level. In all cases this threshold level gives the same volume fraction of graphite: 9% and this value is very close to the real graphite contents in the cast irons under study.

3.4 *Removing artifacts*

Real images often contain some artifacts induced during specimen preparation, like scratches, smearing, relief, pull-outs, comet tails or lapping tracks. Removal of these features is usually very difficult, if not impossible. Simultaneously, any correction of the initial image can affect the features being analyzed and we risk losing control over the whole analysis process. Therefore, every possible effort should be made to get high quality initial images.[5,15,24,43,56,99]

Unfortunately, improvement in the specimen preparation or image acquisition steps is not always possible. For example, it may happen that we get some archival, poor quality images for analysis and there is no possibility of grabbing new ones. Some limitations in specimen preparation are also possible, particularly in the case of smaller laboratories. In such cases we have to improve the image quality before final analysis.

The most common case is a specimen with scratches. A few small scratches can usually be found even in very carefully prepared specimens and these can be relatively easy to remove. For our analysis, however, we will take the image with heavy scratches shown in Fig. 3.3, as in this example the effect of any attempt to remove the scratches is clearly visible. Obviously, images of such low quality should not be used for image analysis.

Application of the Fourier transformation is frequently reported as a successful tool for scratch removal, especially in the case of oriented scratches, i.e., scratches parallel to a given direction. Nevertheless, complete removal of such heavy scratches, as in Fig. 3.3, is not possible. A relatively good result obtained after high-pass filtering is shown in Fig. 3.13 as the upper image. The result seems to be rather unsatisfactory, but closer analysis below shows its real value.

Another method of scratch removal is based on simple morphological operations, which can be found in nearly all the image-processing packages. Application of a simple closing procedure with appropriate size (determined experimentally for a given case) yields good results (lower image in Fig. 3.13). Please note that the image with scratches removed by closing (Fig. 3.13 bottom) looks better than that with scratches removed by FFT (Fig. 3.13 top).

One can also work with software that has built-in procedures for the extraction of lines (for example, the Hough transform). These procedures can be effectively used for removal of scratches, but then force individual analysis as every case can require a slightly different treatment.

Experts in image analysis say that all the problems, including artifacts removal, can be successfully solved, but sometimes will cost enormous amounts of time and energy. Only in the case of possible massive analysis elaboration of a suitable procedure is it worth these efforts. On the contrary, if we have only a few unusual images, the mouse etching method, mentioned in Chapter 2, will be the best choice.

Coming back to our example of a structure with heavy scratches, the final goal is to get a clear, binary image of the black phase. If we threshold the initial image, the result will be as shown in the left image in Fig. 3.14. It is evident that the scratches significantly affect the binarization effect.

Removal of scratches from the binary image is possible, but usually connected with a significant modification of the remaining part of the image. So, a better solution seems to be to threshold a gray level image with partially removed scratches, as shown in Fig. 3.13. This can lead to images with perfectly removed scratches as, for example, the right image in Fig. 3.14.

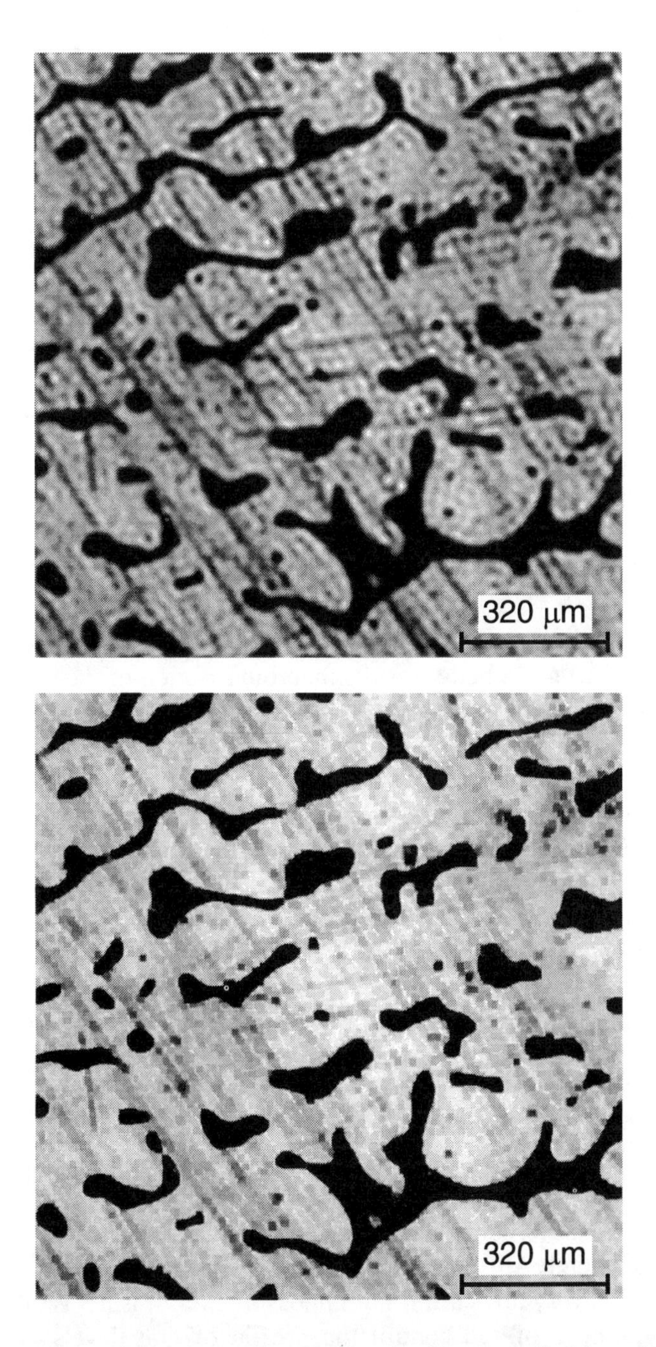

Fig. 3.13. The structure from Fig. 3.3 after partial removal of scratches. Two methods: Fourier transformation (top) and closing (bottom) were used.

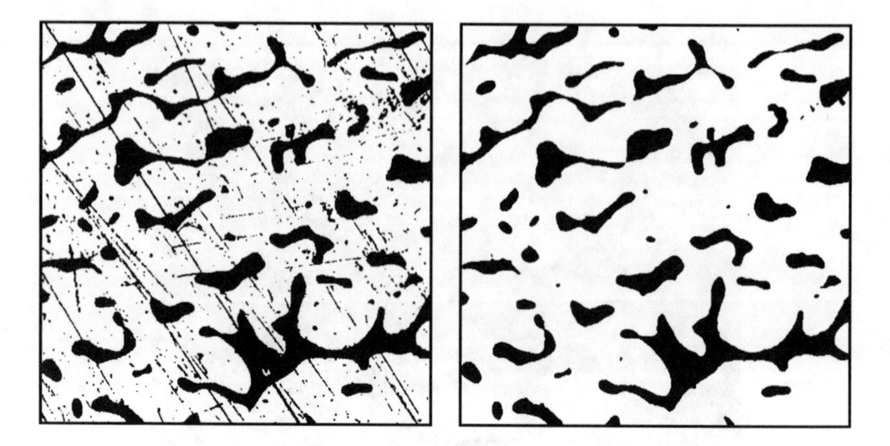

Fig. 3.14. Binary images of the structure from Fig. 3.3: result of a simple thresholding, containing many scratches (left) and clear image with scratches removed, being the result of binarization of the image with scratches removed by appropriate Fourier transformation (right).

It is noteworthy that binarization applied to the image after FFT (Fig. 3.13 top) gives better results than binarization of the image after closing (Fig. 3.13 bottom). So, an image which looks better can be in some cases the worse solution. This seems to be a paradox, but in this case it can be easily explained. In the image after FFT scratches are still visible, but they have a significantly higher gray level than the dark areas to be detected. Therefore, thresholding can effectively remove the scratches from the initial image. Simultaneously, FFT does not affect the shape of these particles as their shape depends on frequencies not filtered out in order to remove the scratches. Thanks to this we get a perfect final binary result, as shown in Fig. 3.14. By contrast, closing offers better removal of the scratches but we have to pay for this result with a more disturbed shape of the particles which we want to detect. Therefore, the quality of the final, binary image is lower in comparison with the FFT method.

The above example illustrates well how difficult any judgment of the choice and fine-tuning of image processing procedures can be. To make things more complex, almost identical results can be obtained using completely different algorithms. Objective evaluation of the result can be very difficult, if not impossible. So, it is advisable to use for the purpose of self-control the profile of gray levels along the given line, as this yields fully quantitative data. Surprisingly, often the best results are obtained after discussion with non-specialists, who are able to look at the solution with a so-called fresh eye. In the case of a search for the solution by an individual, the tips listed below can be helpful.

3.5 Basic tips

It is obviously impossible to cover all the possible cases of image defects and their minimization. Some problems relate only to a given family of images, depending on their origin, specimen preparation, etc. Sometimes finding the satisfactory solution can take years of experience. Fortunately, the analysis of numerous examples allows one to define some guidelines, which can be helpful in finding solutions to individual problems:

- usually it is difficult to perform many tasks that are evident to a human but unclear in a computer program. Therefore, try to define what should be removed or improved. Features for removal can have their unique size, shape, gray level or any other characteristics. Precise definition of these properties can considerably speed-up reaching the final solution
- probably in the majority of cases you will have a lot of difficulties with the removal of anything from the image without alteration of its remaining part. Usually, it is much easier to detect and extract these features than to remove them. Follow this solution and subsequently try to get a suitable result by some arithmetic transformations, mainly subtraction, addition and multiplication
- one of the easiest ways of feature extraction is binarization. Check if the features for removal or modification have their individual range of gray levels. In such a case unwanted features can be easily and safely extracted. In order to check the existence of this case you can use profiles or, even better, several LUT. Many programs offer color LUT, which displays various gray levels as different, often adjustable, colors. This feature enables more precise visual control over the gray levels at distant places within the image
- different sets in the image can possess different attributes. For example, you can detect large and small, round and elongated or bright and dark features. It may happen that features to be removed or analyzed possess properties of various groups and, therefore, are difficult to separate. In such a case you can try to build a few temporary images and reach the final solution using logical transformations. For example, having an image with small particles and a second image with round (of any size) particles, you can use the

logical AND function to create a set of particles that are simultaneously small and round

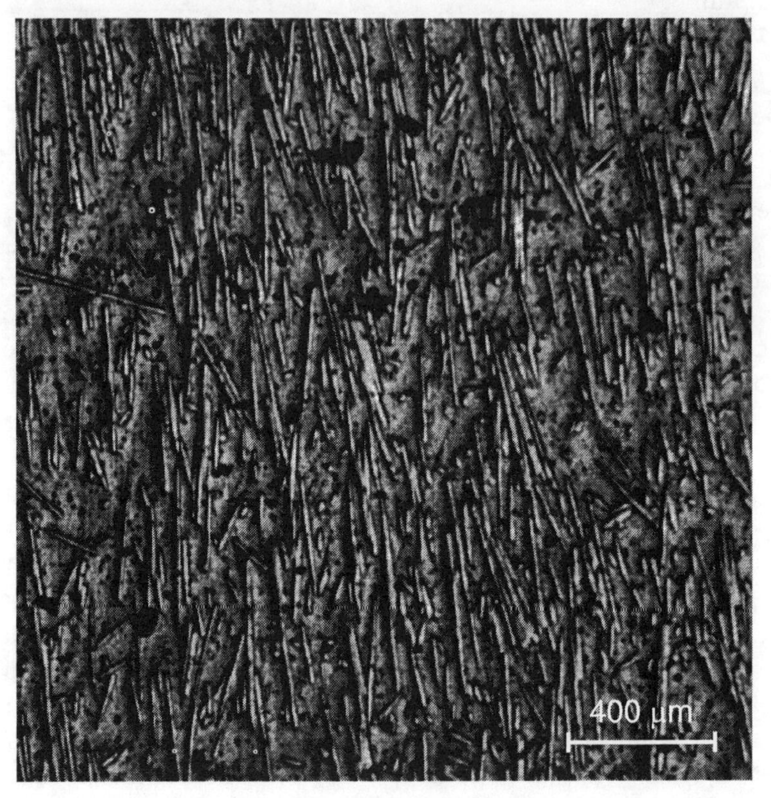

Fig. 3.15. An image can be sharpened with simultaneous contrast enhancement if we multiply the image by itself. The original image is shown in Fig. 3.5.

- if the image has a very poor contrast, you can try to multiply the image by itself. You will improve not only the contrast, but simultaneously the image will appear to be a little sharper. It is a relatively safe operation, because in contrast to classical sharpening algorithms no additional noise is added to the image by the algorithm. An example of such a treated image with low contrast is shown in Fig. 3.15. You can compare it with the original image shown in Fig. 3.5
- in general, any filtering causes the loss of data. In the case of scanned images (this is not applicable to images taken by a CCD camera as they have insufficient size), we can scan them with a resolution two times greater than is necessary for our analysis. For example: if we need to have an image of size 512x512 pixels and

the size of original image is 2x2 inches, we should scan with 256 dpi (dots per inch); doubling this resolution means scanning with 512 dpi and leads to an image of size 1024x1024 pixels. In the image with high resolution we apply all the necessary filters or other operations. Obviously, filters deteriorate the information in the image, but we have an image overloaded with pixels. So, next we should re-sample the image to the desired size (in our example it will change in size from 1024x1024 pixels to 512 to 512 pixels or change in resolution from 512 dpi to 256 dpi). This sub-sampling procedure will compute the new pixel values as the average from four pixels in the high-resolution picture. This trick is very useful for removal of a moiré effect, typical for images scanned from printouts.

If none of these tips works, try to use the ultimate one: simply do another task. Your brain will still work on this problem and switching to another activity gives your brain the time necessary for processing of the problem. A significant number of the problems cited as examples in this book were solved using this technique.

Every user of image processing tools discovers his/her own, unique tips for solving everyday problems. There is no single, universal and absolutely best solution. The tempting beauty of image analysis lies in its flexibility and the unlimited number of possible transformations. Thus, there is always a place to discover tricky, fast and elegant algorithms. In this book you can find some guidelines built from the author's knowledge and experience. These guidelines should help and inspire you, but should never prevent you from your own experiments with image analysis.

Chapter four

Detection of basic features

All the materials: metals, ceramics, plastics, composites or concrete are highly inhomogeneous. In some of them, for example, monocrystals or glass, this inhomogeneity can be observed mainly at the atomic scale; whereas in other cases, with many materials of natural origin, it is visible at a macro-scale as well. Nevertheless, this phenomenon of inhomogeneity is responsible for the creation of the microstructure of any material.

Surely the final user of the material is usually not interested in its microstructure but rather in its properties and price. Unfortunately, the development of clear relations between the material properties and technological processes is too complex to be realized and in practice impossible. However, it was discovered that it is relatively easy to elaborate two complementary sets of relations: the first one between the parameters of technological processes and microstructure and the second one between the microstructure and properties. Consequently, taking into consideration these two groups of relations, the most fundamental rule of materials science can be stated as follows:

> *two materials of identical microstructure have identical properties irrespective of their history, which means the way in which this microstructure was formed.*

It explains why objective, quantitative analysis of the material microstructure is so important in materials science. In this chapter we will focus on the detection of very fundamental features found in the microstructure, whereas the next chapters will deal with different aspects of the more detailed analysis of them.

4.1 Grain boundaries

Grains seem to be the most characteristic feature in the microstructure of a material. They can have different sizes, shapes and characters, starting from clear grains in a clean, solid solution and finishing with very complex grains which are rather inhomogeneous packages of other, similarly oriented, features. To this wide spectrum of grain-type structures one should add the variety of specimen preparation (polishing, etching) and observation techniques (various types of microscopes and observation modes). So, it is absolutely impossible to develop a single, universal algorithm for grain detection and numerous solutions are reported.[3,6,11,13,15,18,23,34,35,56,58,61,65,80,105] Some rules, however, can be established. They will be explained below and followed by a few illustrative examples.

In order to define the grain, we will look for the boundaries surrounding this grain. Therefore, grains and grain boundaries are treated here, from the viewpoint of image analysis, as synonyms.

A general framework for grain boundary detection

The initial image can present various origins and qualities. Therefore, some preliminary treatment may be necessary. There are many possible cases and solutions. Some of them are listed below:

- first, check if the image background is uniform. If not, you can apply any shade correction procedure
- if the image is very noisy, it will be difficult to extract the grain boundaries and some filtering can help in further analysis. This seems to be a good practice to avoid, in this case, sharpening filters. You should try to use filters preserving edges, for example, a median filter. Any filtering should be applied with care, because filters can destroy narrow grain boundaries. Fourier transformation can be a good solution in the case of periodic noise
- thin, dark boundary lines can be effectively thickened by erosion but the whole image cannot be very noisy. Sometimes, better results are achieved using a top-hat transformation
- in good quality images, with a limited amount of noise, the boundary lines can be effectively enhanced using edge detection filters. Unfortunately, this intuitively good solution only occasionally gives correct results
- for some cases the best results can be obtained with the help of user-defined filters. This is, however, an advanced concept which requires some experience. An example illustrating it will be presented later in this section.

The steps listed above can be applied alone or in combination. There exists no universal method of preliminary image processing. Any precisely defined algorithm is valid only for a limited family of images with similar quality, obtained from materials with similar microstructures and specimen preparation. The examples shown in this section should help you to understand how to look for a proper choice of the procedures suitable for a given case.

The preliminary steps are usually followed by binarization, leading to intermediate, binary images. These images can be divided into two groups of entirely different properties, labeled here as Type I and Type II.

Type I. In this case the intermediate binary image represents the continuous network of lines, being a more or less accurate representation of the grain boundaries to be detected. This network can be either a result of single binarization or a logical sum of a family of different images.

This last case requires some more comment: usually, different transformations lead to different detection of the boundary lines (see Fig. 2.18). It may happen that none of these images represents perfect detection, but every one correctly detects a part of the grain boundaries not detected by other images. In such a case summation of these images gives the best possible detection. An example of an intermediate image with a continuous network is shown in Fig. 4.1 (left image).

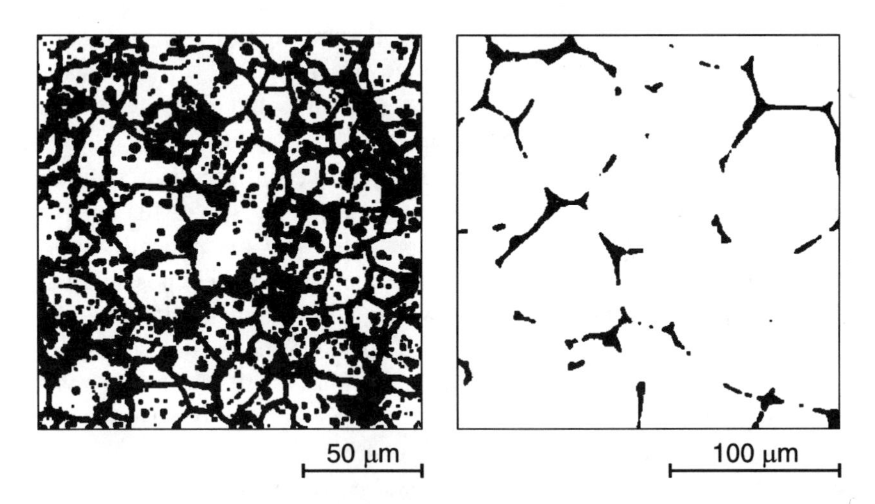

Fig. 4.1. Examples of binary images obtained after preliminary treatment leading to grain boundary detection. They represent both cases, described in the text: case I - continuous network (left) and case II - separated segments of grain boundary regions (right).

Type II. In this event we find the most frequent situation when only discontinuous segments of regions belonging to grain boundaries are detected (right image in Fig. 4.1). This discontinuity can be caused by under-etching, local inhomogeneity of the grains and different sensitivity to etching caused by crystallographic orientation. The other reason can be also high chemical resistance of the material, when the grain boundaries are marked by pores or traces of the eutectic phase only.

Both cases described above require different subsequent treatment and the most typical steps are as follows:

- cleaning of small elements prevents large errors in final boundary detection and can be applied for intermediate images of Types I and II
- in the case of a continuous network (Type I), it is often sufficient to make the network one pixel thin. This can be done using SKIZ
- for a segmented network (Type II), SKIZ will not form the grain boundaries, as pruning (removal of skeleton end-points being the second step of SKIZ) will simply remove the isolated segments. Usually we use the watershed technique (see Section 2.5) for restoration of the lost boundary lines.

Final steps in the analysis can vary from case to case, as will be shown in the examples. Due to the diversity in grain boundary morphology, almost all the possible transformations can be applied in some detection algorithms. The examples presented below will show various possible treatments. A detailed description of all these cases will allow you to trace the detection sequence and find the process route most suitable for your application.

Before switching to the first example, we will analyze an interesting algorithm,[53,76] called *ultimate erosion* (Fig. 4.2). After an erosion of an appropriate size all the particles will disappear from the image. In some cases it is necessary to keep the smallest possible part of the image (this means the part which will disappear after the next step of the erosion). This can be used for counting particles or marking particles glued together. Ultimate erosion is built into many image analysis packages. If you do not find it, you can try to perform it on your own - the idea is illustrated in Fig. 4.2.

A first step depends on erosion and subsequent *reconstruction* of the initial image from the markers, formed during erosion. Reconstruction is a sequence of dilations followed by logical AND with the reconstructed image, repeated until convergence. Reconstruction restores every particle from the initial (reconstructed) image having at least a single point in the marker image. Next, we put the logical difference (XOR operation) between the initial and reconstructed images. This operation yields the set of points (small particles) totally lost during erosion. We repeat the whole loop till the empty set is produced by erosion. In Fig. 4.2 the ultimate eroded set is shown in image *m*. We see from this image that the initial image (*a*) consists of three particles; two of them are glued together.

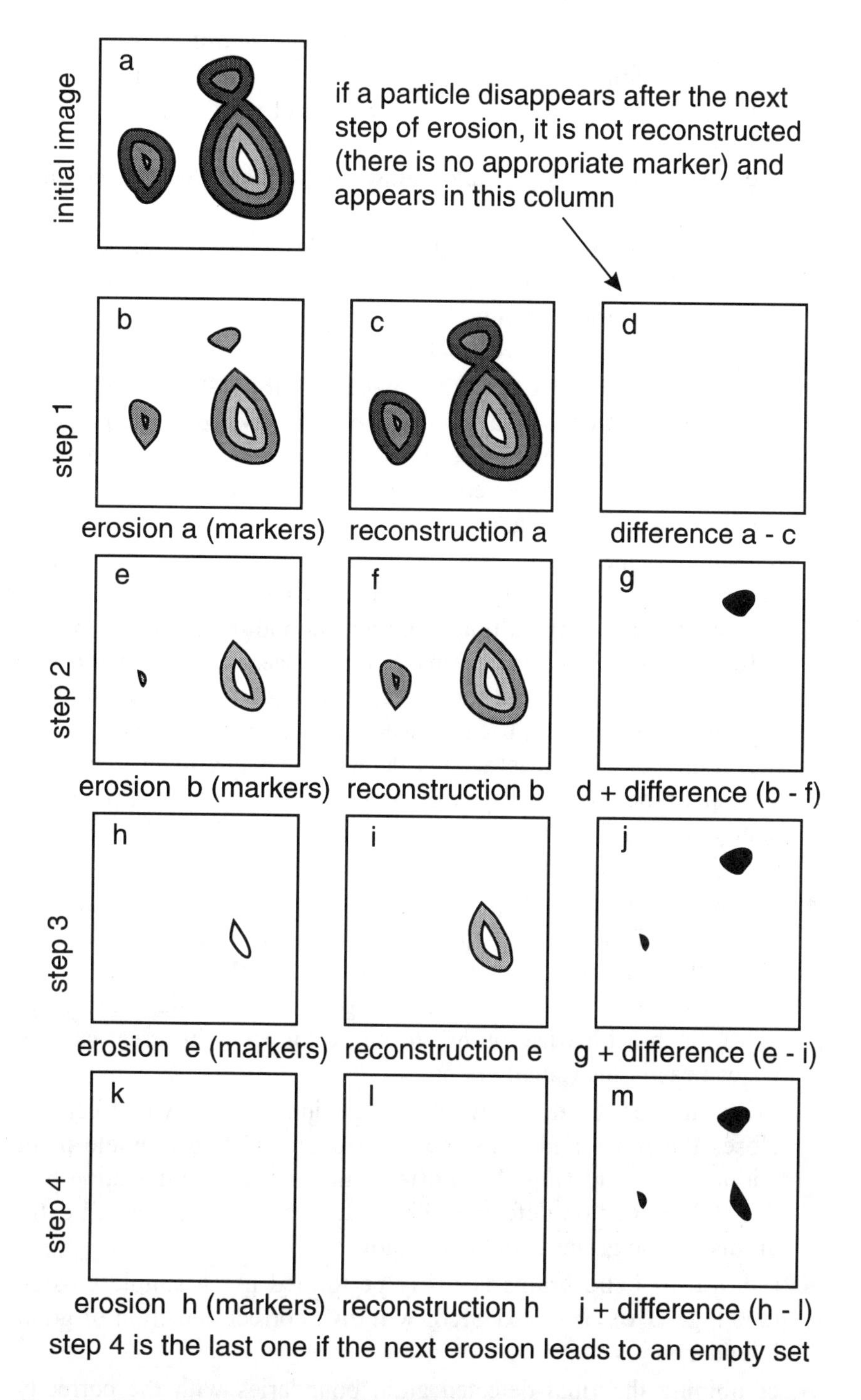

Fig. 4.2. Illustration of the algorithm leading to the ultimate eroded set. See text for details.

Example 1. Grains in a model alloy used for machinability tests

We will start with a relatively easy example - detection of grains in a model alloy, similar to a free machining steel with an elevated volume fraction of inclusions. The main steps of the detection algorithm can be observed in Fig. 4.3. and below you will find its detailed description:

- the initial microstructure is shown in Fig. 4.3a. The background has a uniform gray level, so no shade correction is necessary. As expected, a large amount of inclusions is visible. The grain boundaries seem to be well etched
- the first step in processing the image was the detection of inclusions. Binarization with a suitable threshold level produced a good image of the inclusions. Subsequent opening of size 1 (the smallest possible opening) removed traces of grain boundaries and the smallest inclusions (Fig. 4.3b)
- the smallest inclusions (removed in the step described above) can be preserved if we apply an ultimate erosion. Logical summation (OR operation) of the ultimate eroded set and the structure in Fig. 4.3b will give all the inclusions. You can see the difference in the final Fig. 4.3f, where all the detected inclusions are present. You cannot avoid this step, because when detecting inclusions a part of the grain boundaries is always detected and has to be removed. Thanks to the connectivity with larger inclusions situated predominantly on grain boundaries, traces of unnecessarily detected boundaries are not preserved in the ultimate eroded set
- the next important step is detection of the grain boundaries. A portion of the grain boundaries is relatively bright. So, in order to get the continuous network, we have to apply a relatively high threshold level. This leads to an image with very thick lines (Fig. 4.3c). Such a binarization has to be performed with great care as we can easily lose small grains
- SKIZ applied to the previous image introduces two effects: it closes the holes and thins the grain boundaries to a single pixel width. The result (Fig. 4.3d) offers quite acceptable detection but, due to binarization defects in Fig. 4.3c, we have some small extra grains. These grains should be removed
- removal of extra grains is easily performed using simple erosion (see Fig. 4.3e). The next SKIZ will give correct detection of grain boundaries
- combining the final detected grain boundaries with the correctly detected inclusions (using logical AND) leads to the final effect, illustrated in Fig. 4.3f.

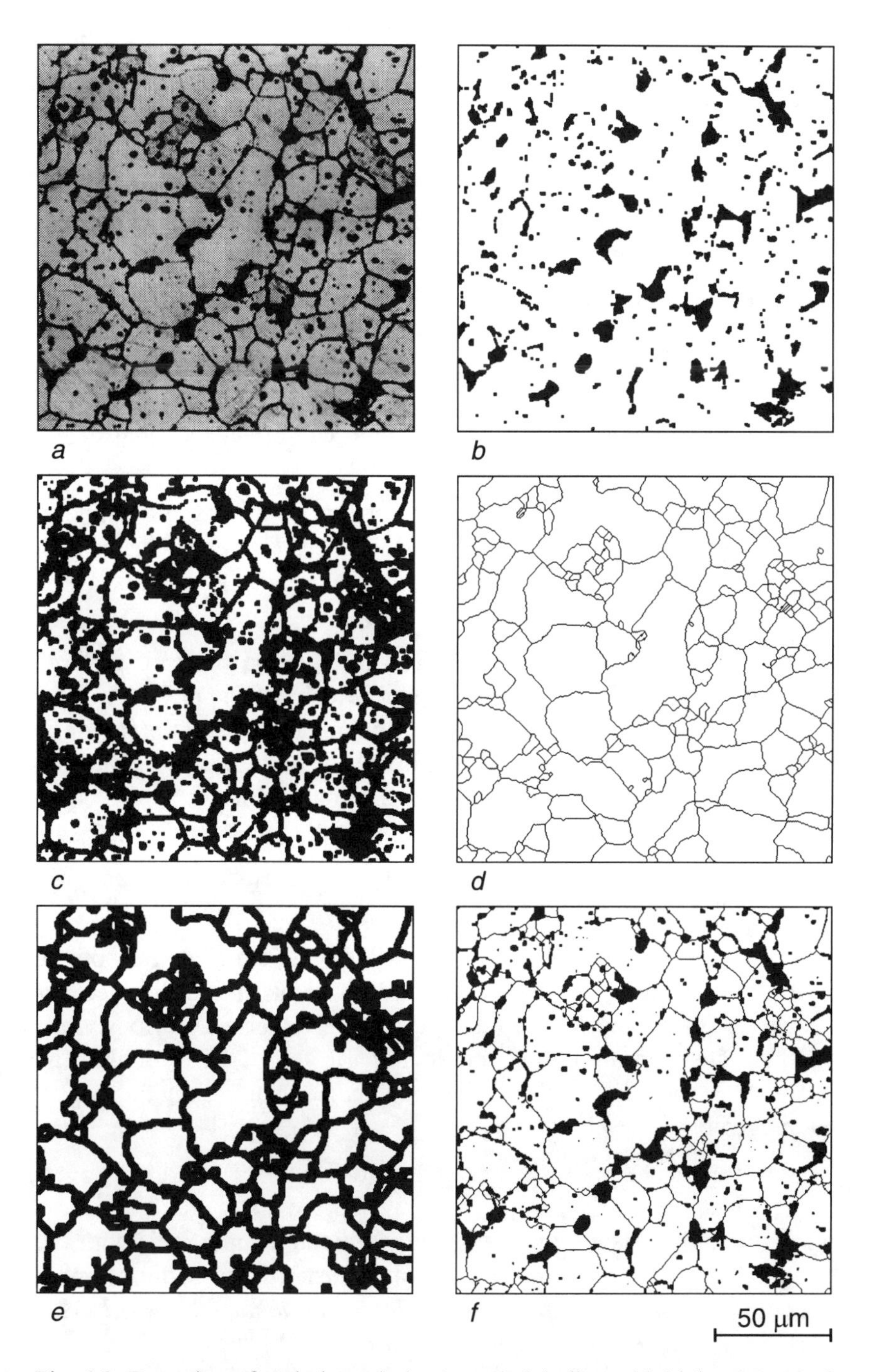

Fig. 4.3. Detection of grain boundaries in a model alloy with high contents of inclusions. Initial image (a), inclusions (b), binary image (c), SKIZ of image c (d), erosion (e) and final detection (f). See text for details.

Note that repeating erosions of growing size and SKIZ continuously removes the smallest grains and produces grains of increasing size. This can be used effectively in algorithms for simulation of grain growth and recrystallization processes.

Example 2. Restoration of grain boundaries in the QE 22 alloy after high-temperature homogenization

The previous example illustrated a typical case of type I grain boundary detection, whereas in this example we will have a typical case of type II detection, with segmented traces of grain boundaries. This example is very important from the instructional point-of-view, as you will see how fine-tuning of the algorithm can affect the final detection.

QE 22 is a very light alloy used for service at elevated temperatures. Its basic components are the following elements: magnesium Mg, silver Ag, zirconium Zr and neodymium Nd. Let us look at the grain boundary detection algorithm:

- in the scanning electron micrograph in Fig. 4.4a you can see large grains (a side effect of the homogenization process) of Mg-Zr solid solution and very bright eutectic Mg-Ag-Nd. The solid solution is inhomogeneous: dark regions are rich in Mg; bright regions, located in the central part of the grains, contain more Zr
- the only possible grain boundary indicators seem to be white precipitates of the eutectic. Its binarization is extremely easy and leads to the next image (Fig. 4.4b)
- the binary image of the eutectic will be used for creation of the distance image (Fig. 4.4d). The darkest points in the distance image are prototypes of the restored grain boundaries. Obviously, for the watershed algorithm one should use the negative of the distance image, as the watershed segments the image along the brightest regions
- unfortunately, watershed segmentation would lead to over-segmentation of the image; we would get too many grains (see how many dark lines are visible in Fig. 4.4d and compare this with the final detection shown in Fig. 4.4e). In order to get better segmentation we will use a *constrained* or *conditional watershed*. This tape of watershed procedure uses not only the distance image but also the markers. This watershed will produce only as many grains as we have markers. So, the choice of markers is, in this case, decisive for the final result

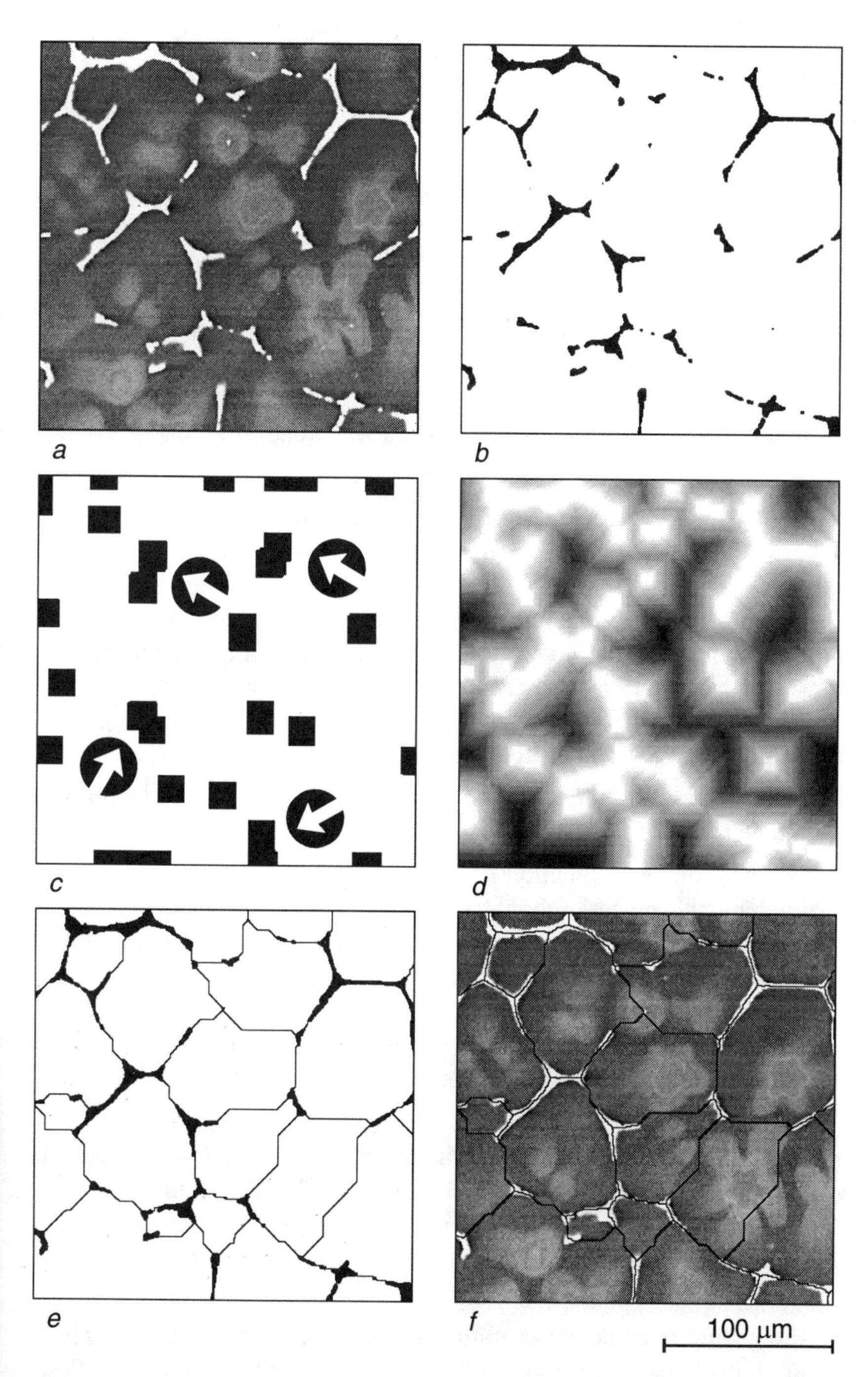

Fig. 4.4. Detection of grain boundaries in the QE 22 alloy. Initial image (a), detected eutectic (b), grain markers (c), distance image (d), final detection (e) and grains overlaid on the initial image (f).

- markers for the constrained watershed can be produced by erosion or, usually better, by ultimate erosion. This gives relatively good results, but due to the discontinuity in the eutectic network we will get too many markers. It could lead to over-segmentation. In order to avoid this effect, we add some steps of dilation - this will glue together markers placed very close to each other. You can observe a few such markers indicated by arrows in Fig. 4.4c
- using the constrained watershed algorithm we can produce the image of grains. Logical intersection (AND) with the image showing the eutectic (Fig. 4.4b) gives the final result (Fig. 4.4e)
- in order to control the quality of our detection we can draw the detected grains as an overlay of the initial image (Fig. 4.4f).

Closer analysis of the above described detection leads to the conclusion that it is somewhat artificial - see unnatural grain boundary lines in the lower right hand corner in Fig. 4.4e. Obviously, this unsatisfactory shape of grains is caused mainly by the poor preliminary detection offered by the eutectic (Fig. 4.4b). Fortunately, we can significantly improve the quality of our detection, taking into consideration the information concerning the e concentration of elements (bright regions in the initial image - Fig. 4.4a). The improved detection algorithm would be as follows (see Fig. 4.5):

- we calculate the distance function as the negative of the initial image (Fig. 4.5a)
- this new distance function has one disadvantage: the regions occupied by the eutectic should be bright, not black as in Fig. 4.5a. Therefore we replace them by the detected eutectic, additionally dilated in order to generalize the image and avoid halo effects at the eutectic-solid solution interface. Combining these two images (i.e., the negative of the initial one and the dilated eutectic) is easily obtained using the arithmetic function that returns the maximum value of the two images (see Fig. 4.5b)
- the final detection is possible using the constrained watershed, in an analogous way to the initial version of the algorithm. Unfortunately, the distance function created from the SEM image forces formation of the watershed as a very curved line and the final detection is not satisfactory (see Fig. 4.5c)
- the geometry of detected grains can be significantly improved after applying erosion (Fig. 4.5d) followed by the SKIZ transformation (Fig. 4.5e). Note that the detected grains are drawn in Figs. 4.5c and 4.5e together with the black eutectic region, whereas for erosion the image 4.5c is taken without the eutectic

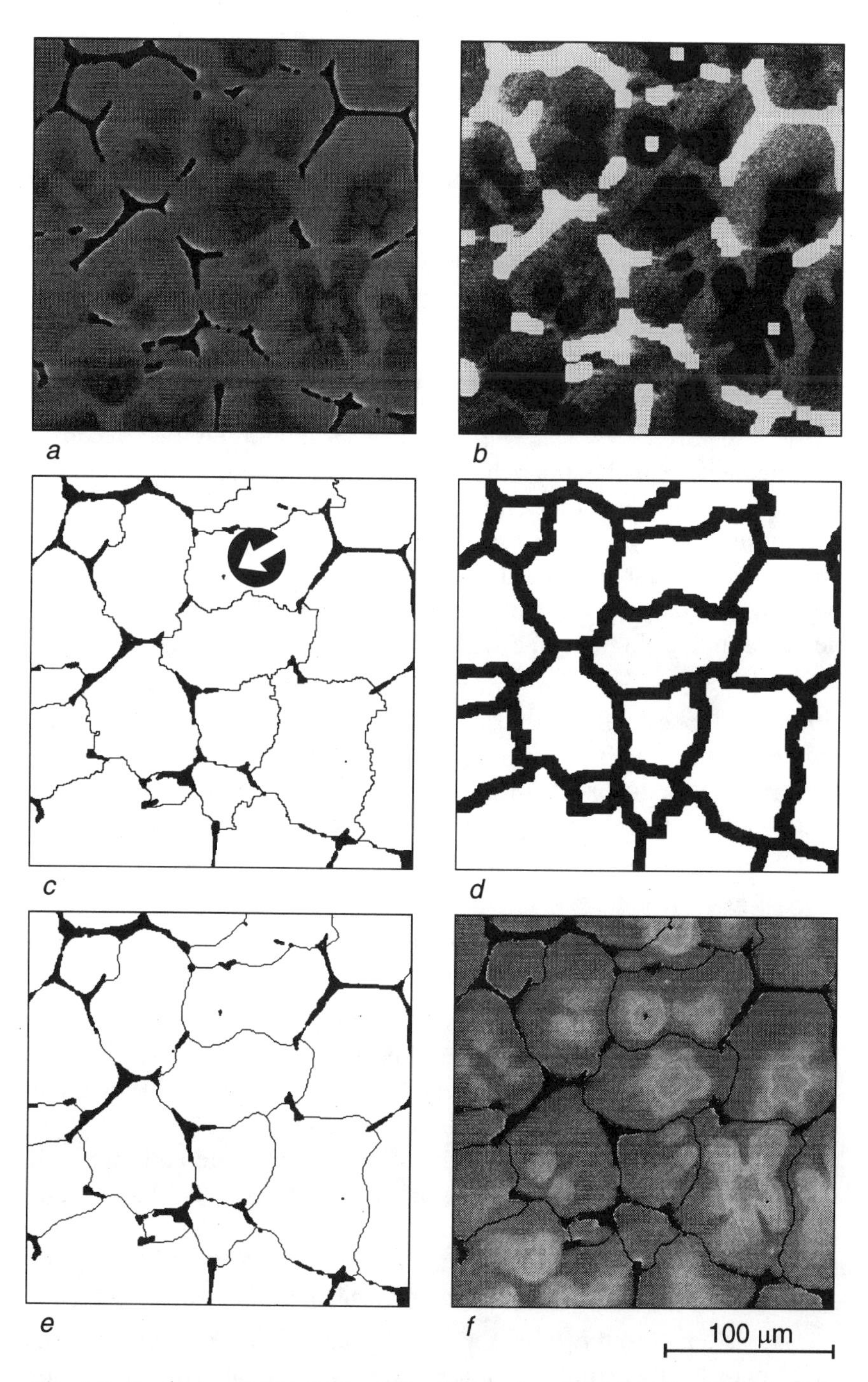

Fig. 4.5. Second variant of detection of grain boundaries in the QE 22 alloy. Negative initial image (a), distance image (b), watershed (c), eroded watershed (d), final detection (e) and grains overlaid on the initial image (f).

- the initial image is plotted in Fig. 4.5f together with the finally detected grains. This allows you to discover the difference between this second variant of the detection and the previous one, shown in Fig. 4.4.

The second detection seems to be significantly better. However, even in this case one can find some errors. For example, there exist a part of the eutectic network (denoted by an arrow in Fig. 4.5c) that is not taken into account during the grain boundary formation process. The previous detection (Fig. 4.4) does not contain this error. This observation illustrates a typical situation in image analysis: we usually have no ideal detection; every algorithm has its advantages and drawbacks. We simply should look for the best possible solution.

In practical applications this is not a critical problem. If we detect correctly the majority of features in a whole series of images, there is a great probability that our error of detection is systematic and has an approximately stable value. In such conditions we can precisely and objectively compare the materials, in spite of the inaccuracy of the detection in the single image.

The last example also shows that any firm classification of algorithms is impossible. We started with a watershed, a typical tool for the detection of grain in structures of type II and the final tuning was performed with the use of SKIZ - a classical tool for processing type I structures. The next example will show even more clearly that the number of possible algorithms is really unlimited.

Example 3. Grains in a polystyrene foam (a rather difficult case)

Polystyrene foam is widely used as a material for heat insulation. In some cases it also has to carry some mechanical forces. Both strength and insulation properties are controlled by the form and size of the foam cells. The cells in polystyrene foam are clearly visible without any microscope, as their largest chords are a few millimeters long (see Fig. 4.6a). Simultaneously, the whole material is white and therefore the contrast in its image is very poor. A good solution could be rotating the side light source and then taking a few images of the same region, but illuminated in different ways. However, here we will try to develop a suitable algorithm based on the analysis of a single image.

Any attempt to threshold the initial image of the polystyrene foam gives poor results, similar to that shown in Fig. 4.6b. The human observer can guess that there are some grains in the image, but it is too difficult a task for a computer. Therefore, we will try to develop a new algorithm that will be described below:

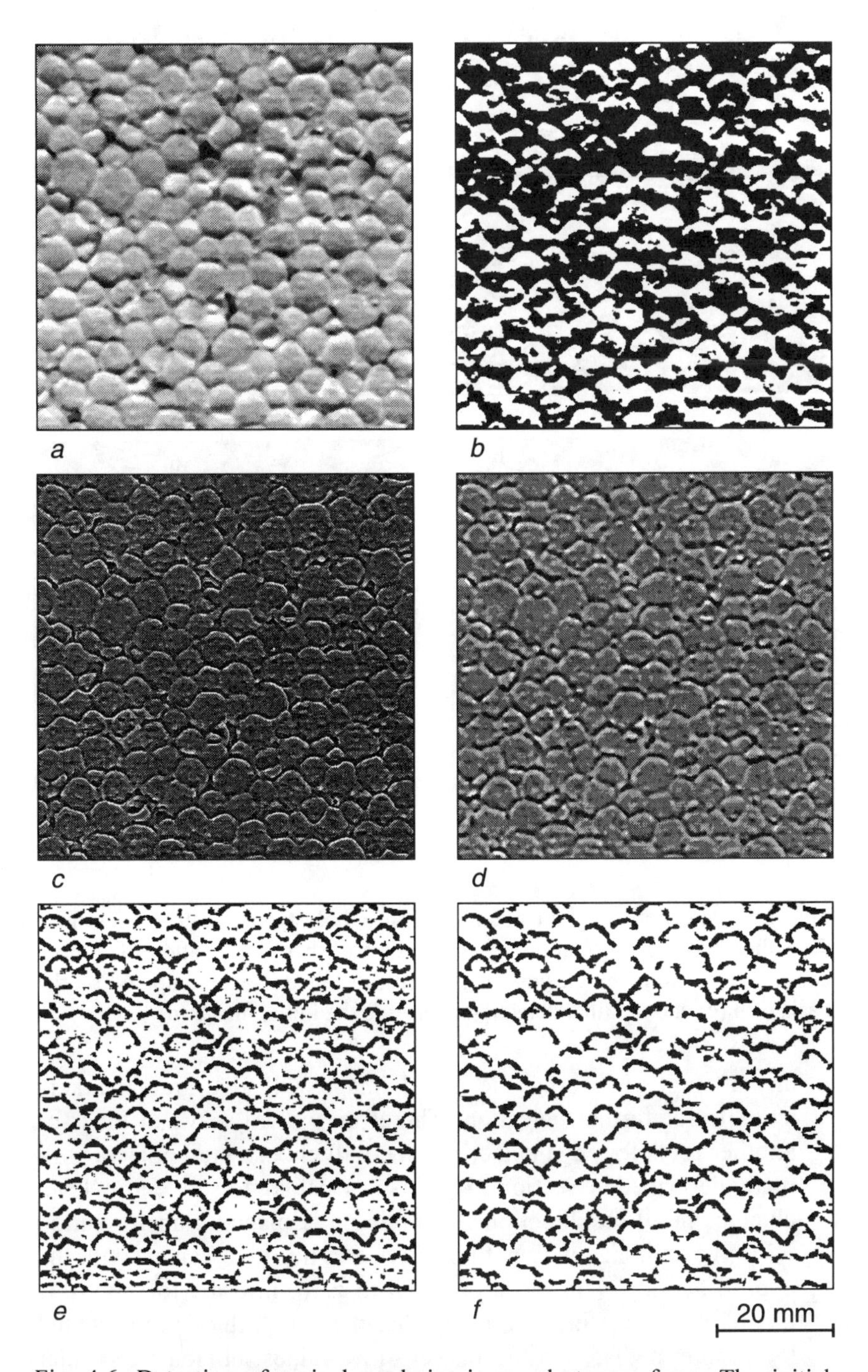

Fig. 4.6. Detection of grain boundaries in a polystyrene foam. The initial image (a), binarized a (b), Laplacian image (c), Laplacian smoothed by median filter (d), thresholded d (e) and e with removed small features (f).

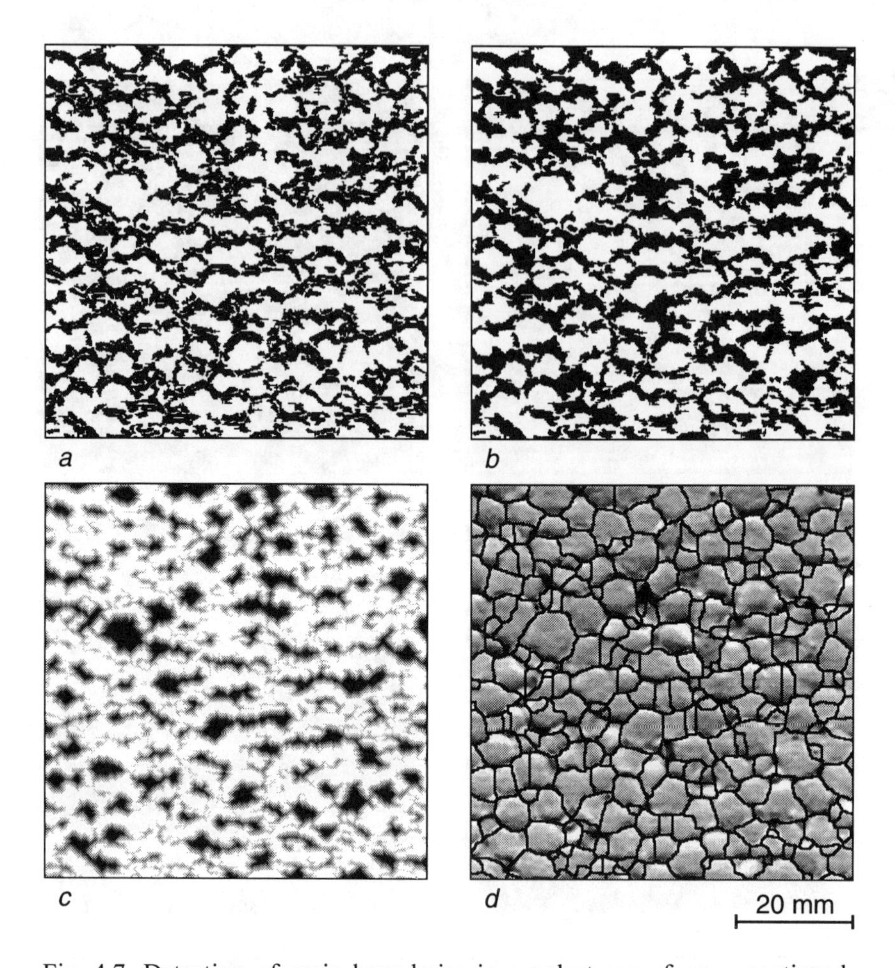

Fig. 4.7. Detection of grain boundaries in a polystyrene foam - continued. Two thresholds of the Laplacian combined together (a), creation of a more continuous network by closing image a (b), distance image computed on the basis of b (c) and the final detection overlying the initial image (d).

- any attempt to binarize the initial image will give grain markers suitable for further detection. Therefore we will have to look for another tools for edge detection. We will choose the Laplacian, already described in Section 2.7. This is sensitive to subtle gray level changes in all the directions. Applying a Laplacian transformation will give an image with both positive and negative values, with the overall average very close or equal to zero. In order to show this image, we plot it in such a way that the zero value is represented by an intermediate gray level, very low values (obviously, smaller than zero) are black and very high values are shown in white (Fig. 4.6c)

- due to the low contrast of the initial image, the Laplacian is too noisy for further treatment. In order to improve this image we will use the median filter, which limits the noise level and simultaneously preserves the edges (see Fig. 4.6d)
- the filtered Laplacian can be used for preliminary extraction of the grain boundaries, visible in Fig. 4.6d as very bright or very dark areas. We will extract the bright component by an appropriate threshold. This creates the binary image, shown in Fig. 4.6e
- some remnants of the noise are still present in the Laplacian, even after median filtering. Therefore, in the binary image of the bright component (Fig. 4.6e), many small particles can be observed. They can be removed on the basis of surface area analysis - we simple remove the features with the smallest surface area (see Fig. 4.6f). If there is no suitable procedure in your package, you can try to use opening, but the result will be not as good (opening is not sensitive to the surface area but to the breadth of the feature - it will remove long, thin features even it they have a large surface area)
- an analogous binary image can be obtained after thresholding of the dark component in the filtered Laplacian. This intermediate image is not shown in the illustrations
- the next step is combining both intermediate binary images, using the logical OR (Fig. 4.7a). As a result, we get a series of approximately horizontal clusters. The lack of a vertical component is caused by the directional illumination of the specimen (see Fig. 4.6a)
- the detected traces of regions close to the foam cell boundaries are highly discontinuous. A simple closing produces a significantly more continuous network (Fig. 4.7b), which can be used to form the distance image
- as usual, for the needs of restoration of grain boundaries we will use the negative of the distance image (refer to Section 2.5). This form of the distance image is presented in Fig. 4.7c
- the distance image allows us to detect the boundary lines using the watershed transformation. The binary image used as the starting point for computing the distance function was smoothed using the closing procedure (Fig. 4.7b) and, therefore, an ordinary watershed can be applied (a constrained watershed is unnecessary in this case). It simplifies the algorithm and leads to the final detection. The detected grain boundaries are overlaid over the initial image and shown in Fig. 4.7d.

Careful analysis of the final result (Fig. 4.7d) shows that approximately 5% of grains are over-segmented, this means additionally divided into two pieces. This does not change the overall rating of good segmentation - one has to take into account the low contrast of the initial image, making automatic image processing very difficult.

Example 4. Grains of a clean, single phase material

The next example illustrates the classical problem of grain boundary detection in a single-phase material - in our case it is ferrite after recrystallization. The proposed algorithm for grain boundary detection consists of the following steps:

- the initial image offers well-etched grain boundaries and at a first glance it should not be particularly difficult to detect them (Fig. 4.8a)

- grain boundaries are easily visible to a human, but they are very delicate and some shading makes the final detection difficult. A very good detection can be obtained with the help of a black top-hat transformation (Fig. 4.8b)

- the next steps in the algorithm are erosion, leading to a continuous network and hole filling, removing small, isolated particles. The result of these operations is shown in Fig. 4.8c. Hole filling is often built into image analysis packages. Its principle is based on one of the most common functions in computer graphics, namely flood filling. This operation fills all the regions connected with the image boundary. In our case it would be the continuous network formed during erosion. XOR with the initial image returns all the holes and a second XOR with the initial image restores the continuous network, this time without holes

- first detection of the grain boundaries can be obtained after SKIZ is applied to the image in Fig. 4.8c. Unfortunately, this detection will not detect all of grain boundaries. Some of the missing boundaries are indicated by arrows (Fig. 4.8d)

- some improvement of this detection can be obtained with the help of a watershed detection. On the basis of roughly detected grain boundaries (Fig. 4.8c), we build the distance image (its negative is shown in Fig. 4.8e) and the final detection is a result of the watershed detection (Fig. 4.8f). This second detection is better, but we have still some over-segmentation or lost boundary lines. Two examples of these errors are indicated by arrows in Fig. 4.8f.

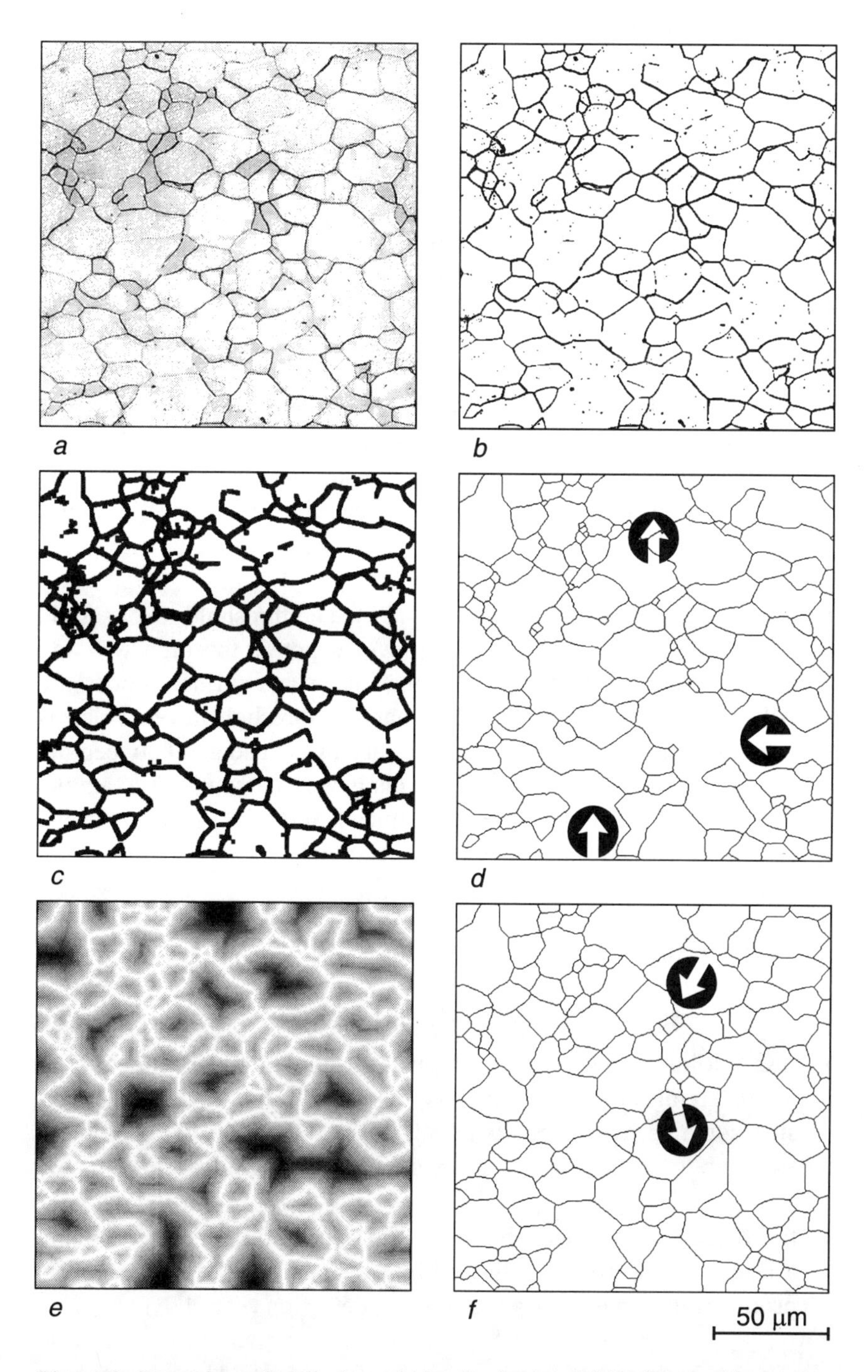

Fig. 4.8. Detection of grain boundaries in ferrite. Initial image (a), top-hat transformation and detected features (b), removal of small parts and continuous network (c), SKIZ (d), distance image (e) and final detection (f).

In this example of grain boundary detection we face one of the very common problems in image analysis. We observe quite clearly some features that are extremely difficult for automatic detection - these are the lost boundary lines. So, we would like to get better detection.

For testing the appropriate algorithm we will take a part of the initial image, containing some boundary lines lost during further detection. This part is shown in Fig. 4.9a. Next, we apply the following transformations:

- some of the boundary lines exhibit very low contrast with the matrix and any attempt to threshold them fails to detect them because of the noise. A simple opening transformation will reduce this noise. The result is very difficult to discern by a human (Fig. 4.9b) but is sufficient to improve the detection

- the next step is binarization (Fig. 4.9c). The threshold level should be carefully chosen. A threshold level which is too high will add much of the low-level noise to the image and further processing will be impossible. On the other hand, a threshold level which is too low will not detect all the boundary lines. The gap between the too low and too high threshold levels is very narrow. For example, in Fig. 4.9b a part of the detected lines is too thick, whereas in the other part some discontinuities are observed

- in order to remove the above-mentioned discontinuities we apply erosion (Fig. 4.9c), leading to a continuous network around all the grains

- SKIZ following erosion will give us thin boundary lines (Fig. 4.9e), giving almost satisfactory detection. This image, however, still contains a few small defects, like a single, additional, very small grain or some unnecessary steps on the boundary lines

- the defects mentioned above can be successfully removed using a few steps of erosion followed by a second SKIZ operation. This produces the final, very good quality, detection (Fig. 4.9f).

So, we have an algorithm giving perfect results. Unfortunately, it cannot be applied to the whole image. The next figure (Fig. 4.10) will explain the reason for this.

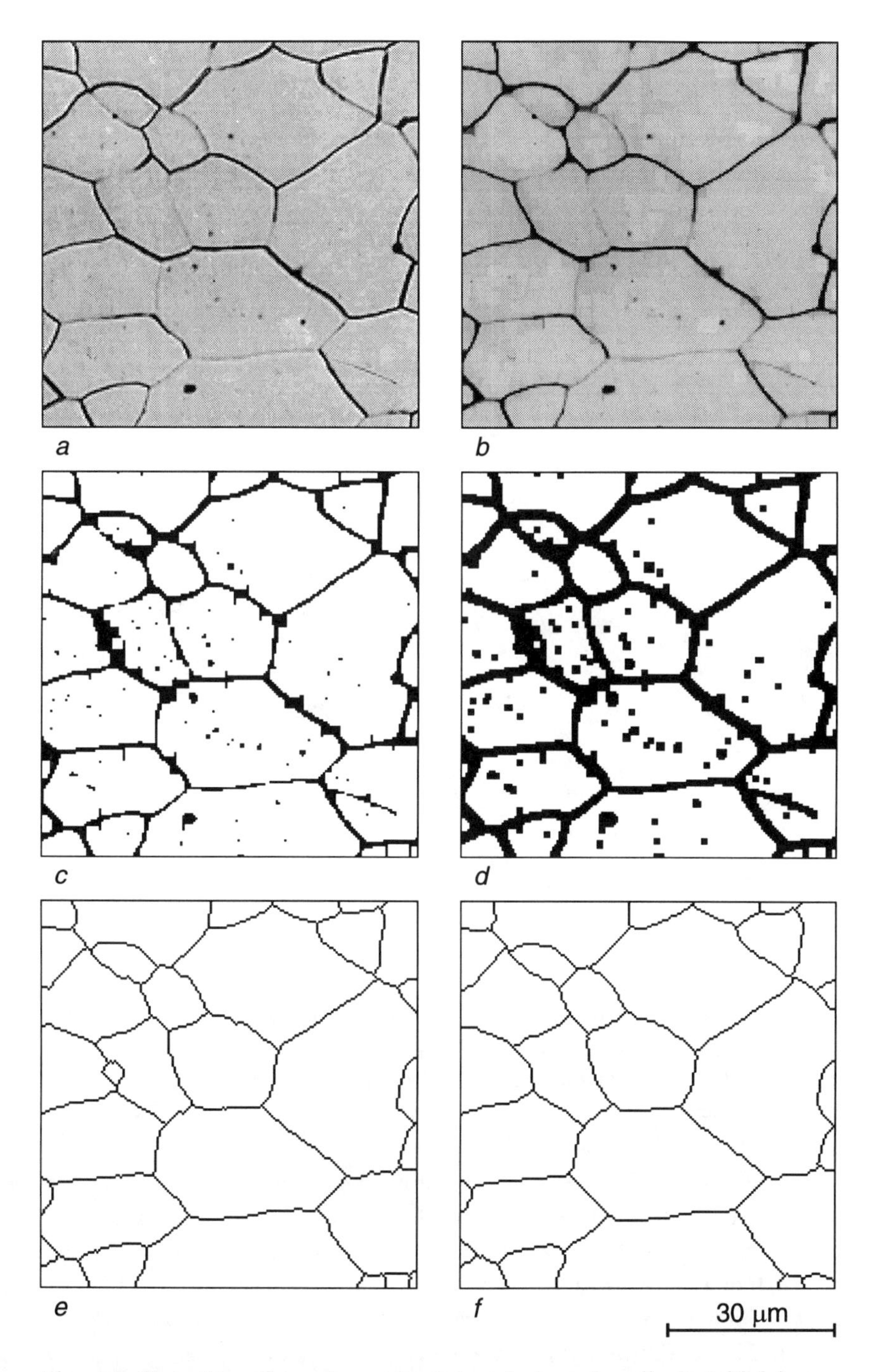

Fig. 4.9. Detection of poorly contrasted grain boundary lines. Initial image (a), opening (b), binarization (c), erosion (d), SKIZ (e) and erosion followed by the second SKIZ (f).

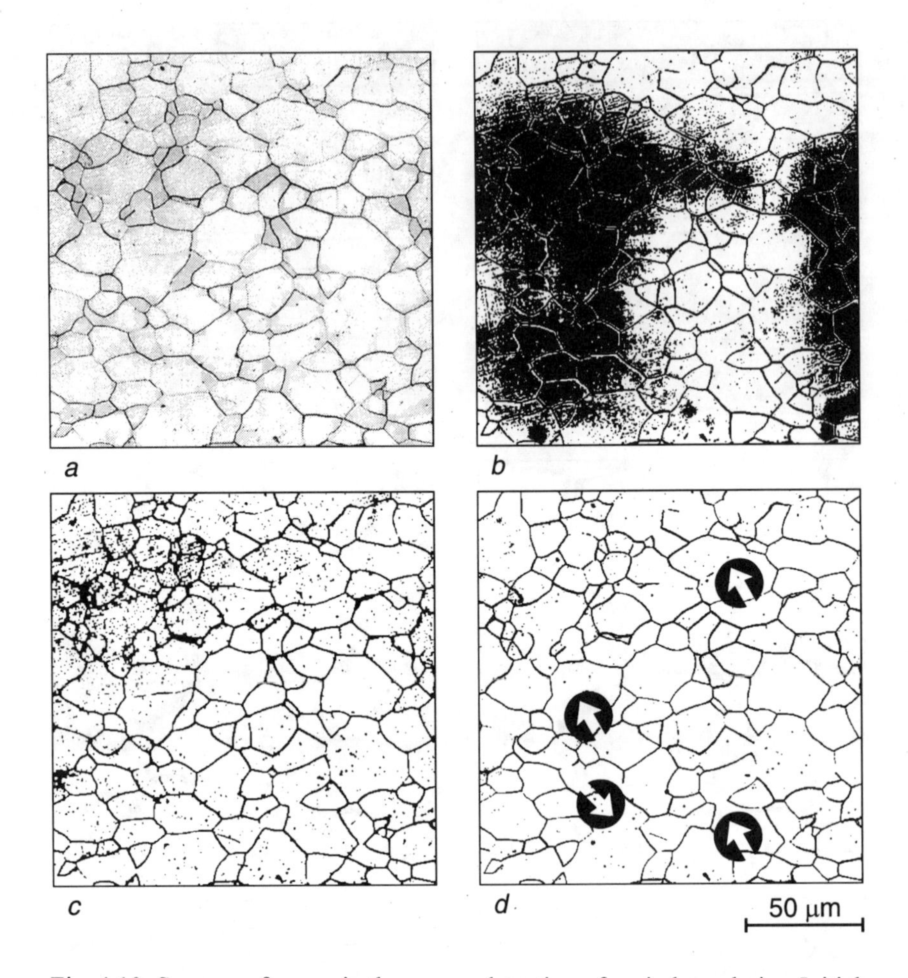

Fig. 4.10. Sources of errors in the proper detection of grain boundaries. Initial image (a), thresholded image distorted by shading (b), thresholded image after carefully removing shade - some noise is still visible in the upper left-hand corner of the image (c) and the optimally thresholded image, with some lost boundary lines (d).

Any attempt to threshold the initial image will lead to biased detection. It is a straightforward effect of the shading present in the image (Fig. 4.10a and b). Careful correction of the shading will remove this effect but shade correction is unable to correct the local noise present in the upper left-hand corner of the image (Fig. 4.10c). Application of the optimal threshold level will not detect the noise, but some of the boundary lines will be lost. Some selected regions of the lost boundary lines are indicated by arrows in Fig. 4.10d.

So, perfect detection is possible in this case only in sub-regions of the initial image. We can perform this procedure in every sub-region and then build the final image from the partial ones. However, as will be shown in Chapter six, devoted to measurements, it should be stressed that small errors in automatic detection can have a negligible effect on the final results of measurements.

Example 5: Grains in a CeO_2 ceramic

This is the next example of grain boundaries that are clear to a human and simultaneously extremely difficult for extraction using computerized tools.

The initial image is shown in Fig. 4.11, together with the profile plot, illustrating gray level values along the horizontal white line situated in the center of the initial image. It may be surprising that the plot is so flat, but you should note that the brightness and contrast of the initial image were improved for the needs of presentation. In the next figures, starting from Fig. 4.12, the initial image will be shown without any modification. It is worth noticing that any linear transformation (brightness and contrast control belong to this family - see Fig. 2.1) will not affect the difficulty of subsequent analysis as all the pixel values within the image change proportionally.

Analysis of the plot in Fig. 4.11 shows that only the first (from the left) crossing of a grain boundary gives a clear response in the curve. This high peak, however, arises from the presence of a pore in the vicinity of this grain boundary. The remaining grain boundaries give peaks of a size that can be compared with the local noise level. So, it will be really difficult to detect the grain boundaries and we will trace in detail the applied algorithm:

- Any attempt to get the traces of grain boundaries using the binarization procedures will fail due to the low contrast in the initial image (Fig. 4.12a)
- the effect of a grain boundary can be compared with the local noise (see Fig. 4.11) but, fortunately, these boundary points are placed in some order and therefore we can try to apply some edge detection filters. We will apply the Prewitt filter (see Section 3.5). It takes information from a 3x3 pixel area and therefore should minimize the effect of local noise. Detection of grain boundaries using this filter gives no continuous lines but seems to be quite satisfactory (Fig. 4.12b)

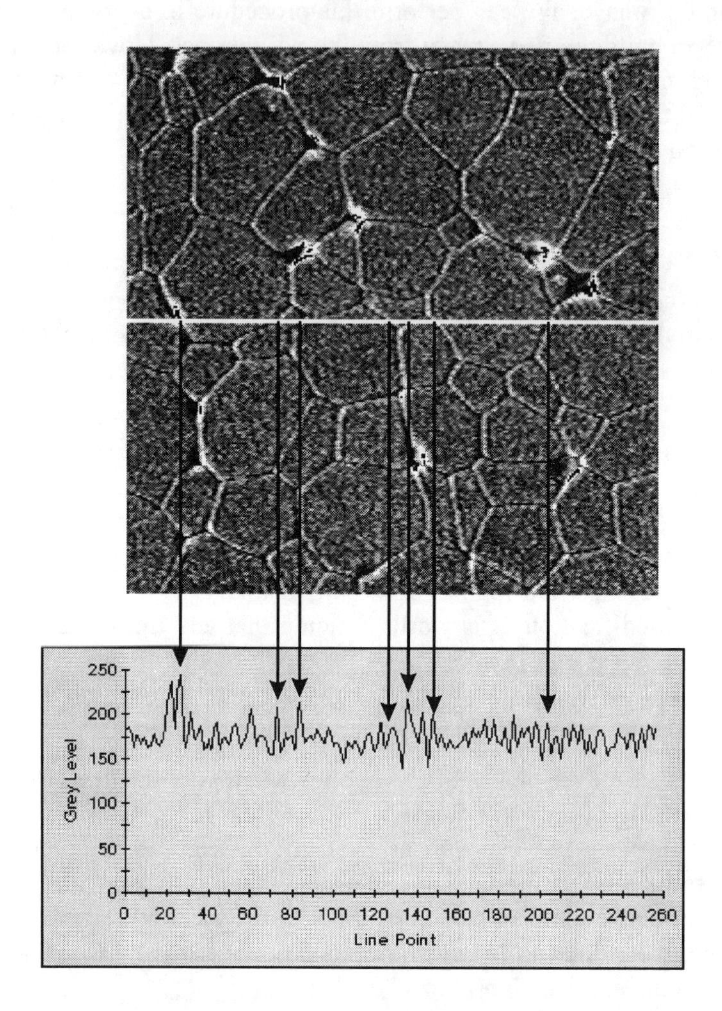

Fig. 4.11. Initial image together with the gray level plot along the white, horizontal line in the middle of the structure. Note that, in general, grain boundaries give no significant peaks suitable for grain boundary detection.

- the image produced by the Prewitt operator is very noisy. We should remove this noise; and the median filter seems to be a good solution (Fig. 4.12c)
- in order to produce the relatively continuous network we will apply erosion to the image after median filtering. Too many steps of erosion would likely destroy the subtle structure of grain boundaries, so you should apply only a small amount of erosion (Fig. 4.12d)

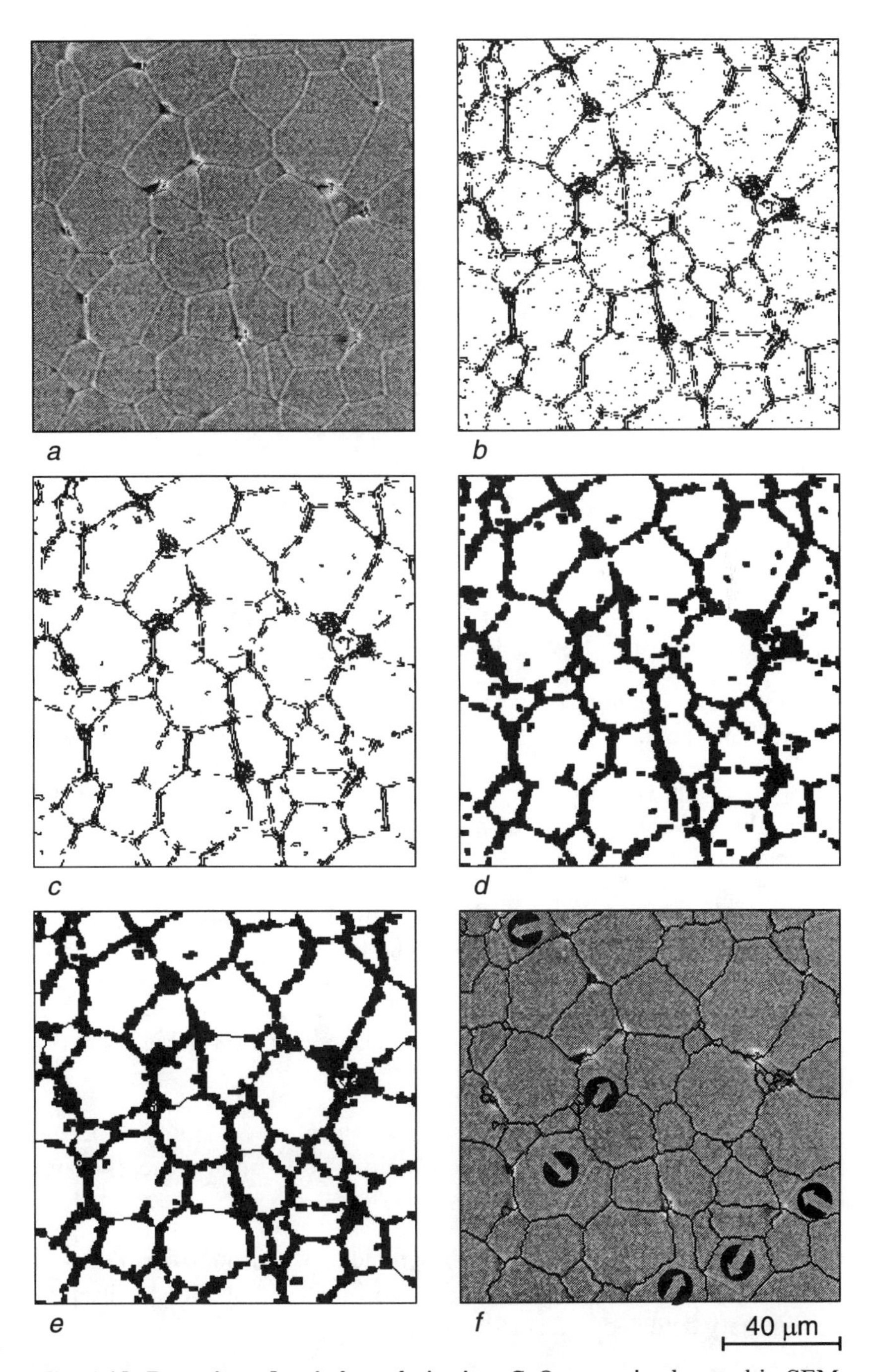

Fig. 4.12. Detection of grain boundaries in a CeO_2 ceramic observed in SEM. Initial image (a), Prewitt edges (b), edges median filtered (c) and eroded (d), hole filling and watershed detection (e) and final detection overlaid over the initial image (f). Arrows indicate faults in detection.

- the eroded image presents a quasi-continuous network with some holes. In the next step we will close the holes and detect the missing grain boundaries using the classical watershed approach. The result of these transformations is shown in Fig. 4.12e

- the last step in our image processing will be a SKIZ that will produce the final, thin network of detected boundary lines (Fig. 4.12f). The result of detection seems to be satisfactory - the detected grain boundaries are located at the correct places, which can be verified in their image overlaid over the initial image. In spite of a good overall result, some boundary lines are not detected (arrows indicate these places). Therefore, we will look for a more accurate detection procedure.

In this new procedure we will try first of all to improve the introductory detection. It is rather difficult to find a single step procedure offering ideal detection. Thus, we will combine three intermediate images obtained using different detection procedures. The whole procedure will look as follows:

- first, we will detect from the initial image (Fig. 4.13a) the Prewitt gradient image (Fig. 4.13b), exactly, as in the previous algorithm

- next, we apply a white top-hat transformation in order to detect the bright halo effect observed at the grain boundaries thanks to the local curvature of the specimen (Fig. 4.13c)

- in a fully similar way we compute the black top-hat (Fig. 4.13d) that reveals the darkest regions in the initial image

- using the logical AND we build the image with significantly better detected boundary regions (Fig. 4.13e), which are the basis for the next transformations

- as in the previous algorithm, we will clean out the noise using the median filter (Fig. 4.13f). We obtain a good quality, nearly continuous network

- further improvement is done by erosion, which improves network continuity and fills holes, leading to improved quality of the preliminary detected grains (Fig. 4.14a)

- the detected network is now relatively thick and the grains have very irregular shapes. We can easily improve the shape of the grains by simple dilation. This image now requires some fine-tuning in order to restore the missing boundaries (Fig. 4.14b)

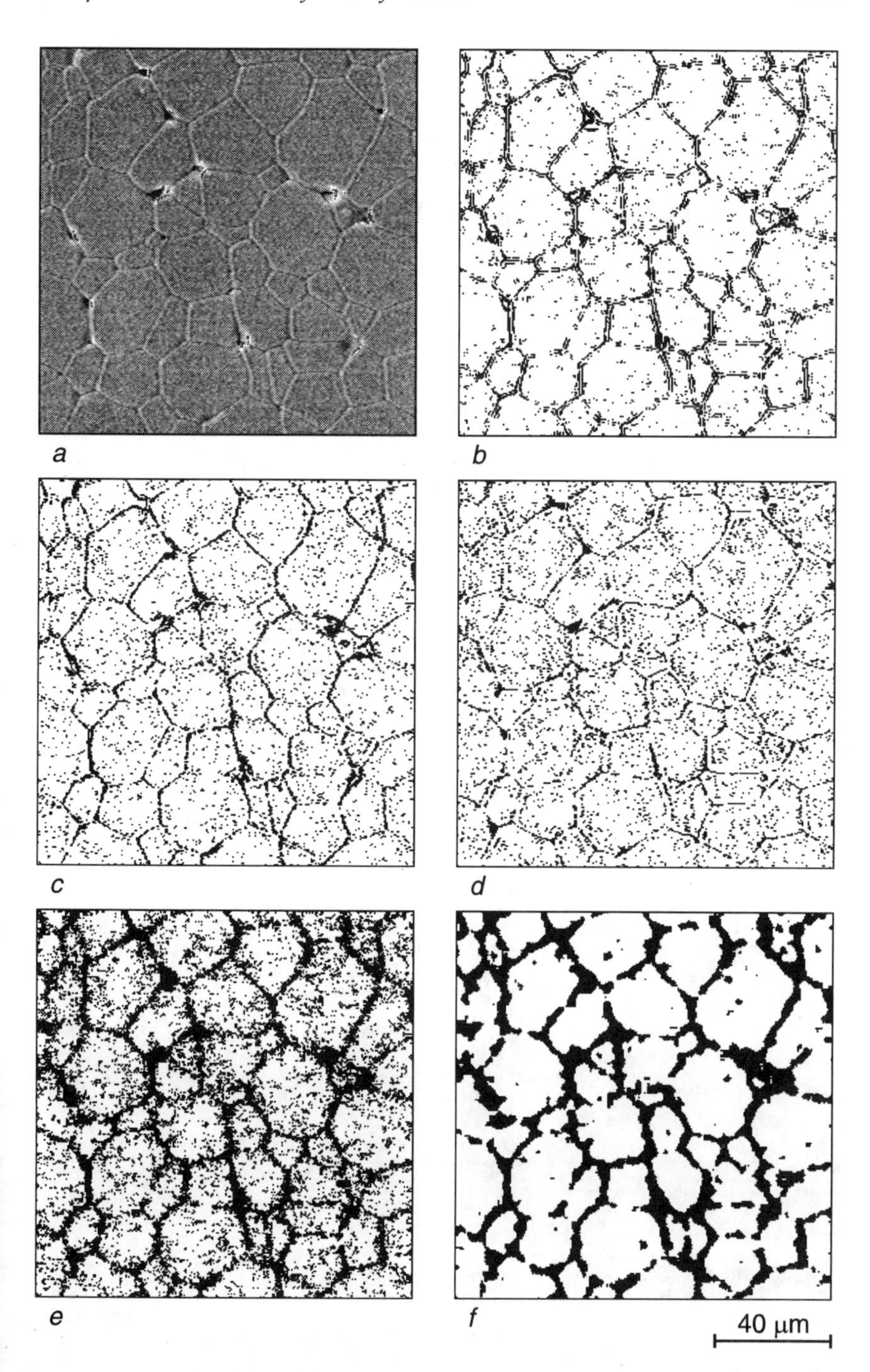

Fig. 4.13. Second algorithm for the detection of grain boundaries in a CeO_2 ceramic. Initial image (a), Prewitt edges (b), white top-hat (c), black top-hat (d), images b, c and d combined together (e) and the compound image after noise removal by a median filter (f).

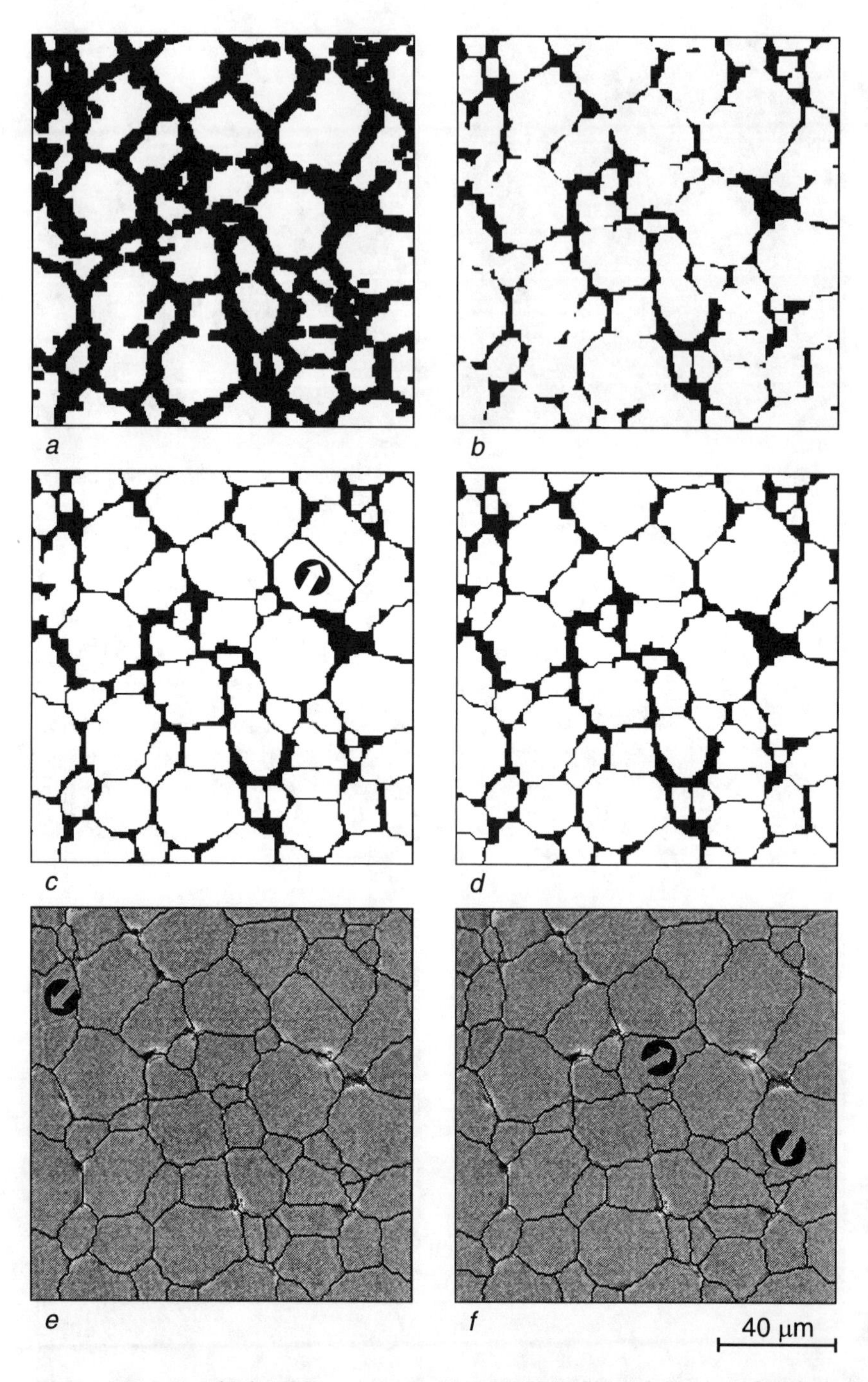

Fig. 4.14. Second algorithm - continued. Median filtered image after hole filling and erosion (a), dilated image (b), 4-connected watershed (c), 8-connected watershed (d), SKIZ of image c (e) and SKIZ of image d (f). Arrows indicate erroneous detection regions.

- in order to restore the lost boundary lines we will apply, as usual, watershed detection. We will do it in two ways: 8-connected (Fig. 4.14c) and 4-connected (Fig. 4.14d). You can easily see the difference: 8-connected watershed produces an extra line (indicated by an arrow)
- the final step will be a SKIZ (in both cases in the 8-connected mode), which leads to the final detection, shown in Figs. 4.14e (8-connected watershed) and 4.14f (4-connected watershed)

Some remarks concerning the difference between both of the used 4- and 8-connected modes are given later in this section. Here, we will discuss only the quality of the final detection. In the case of the detection based on the 8-connected watershed (Fig. 4.14e) we have an extra line and have totally lost one crossing of the boundary lines, visible in the upper left-hand region of the image (indicated by an arrow). On the other hand, detection based on the 4-connected watershed (Fig. 4.14f) lost two short segments, also indicated by arrows. In the overall rating, the 4-connected watershed gave better detection here than the 8-connected watershed. However, the difference is really small and both methods of detection offer significantly better results than the first algorithm (see Fig. 4.12f).

From this last example we can draw some conclusions that are important for practical applications:

- there are numerous solutions, leading to similar results. It is almost impossible to judge, what algorithm is the best one - it can depend on many variables, like the hardware and software available, quality of initial images, features to be detected, etc.
- minimizing the detection inadequacy usually requires the use of more complex algorithms that are less flexible and difficult to control in the case of massive data flow.

Now we will switch to the promised remarks on connectivity. *Connectivity*[53] is a property of the binary image components allowing the computer to decide if the analyzed features are separate or form a single, *connected* component. We will not analyze here the whole theory; we will discuss just the properties which are most important for practical applications. As was mentioned in Chapter 2, we can find two basic matrices of pixels in image analysis: square and hexagonal. In the case of the hexagonal grid, there is no problem: if two points touch each other, they produce unity, i.e., are connected. Unfortunately, only a limited number of image analysis devices use the hexagonal grid. Most of them, following the graphics adapters in computers, use a square grid of pixels. All the algorithms analyzed in this book were processed with a square grid.

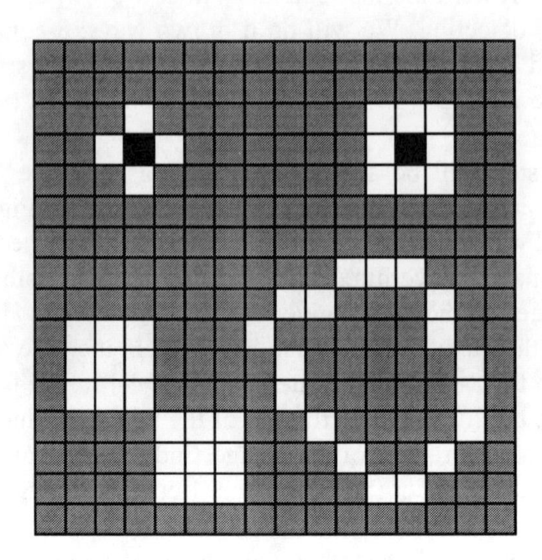

Fig. 4.15. Illustration of the connectivity in square grid. See text for details.

In case of the square grid we can define the number of closest neighbors of any pixel. If we assume that a pixel has 4 closest neighbors (see four white pixels surrounding the black one in the upper left-hand part of Fig. 4.15), we perform 4-connected analysis. If we decide that any pixel (see the upper right-hand corner in Fig. 4.15) has 8 closest neighbors, we perform 8-connected analysis.

Changes in connectivity can seriously affect our analysis (one example, showing minor changes is shown in Fig. 4.14). For example, two squares visible in Fig. 4.15 form two objects if treated as 4-connected but they form a single figure if treated as 8-connected. An even more dramatic effect will be for the curve plotted in the lower right-hand corner in Fig. 4.15. This will be considered as a curve in 8-connected space, whereas in the case of 4-connected analysis it will be considered as 10 separate particles.

So, in some cases an 8-connected analysis takes some isolated features as one. This is especially important if we start measurements in images after, for example, a SKIZ operation, as the particles are separated in this case only by a one-pixel-width line. On the other hand (note the example curve in Fig. 4.15), operations connected with particle growth, hole filling, etc., usually return better results if performed as 8-connected. Obviously, the decision about connectivity has to be made individually in each case, but it is worth knowing how it can affect the results of our analysis.

Example 6: WC-Co cermet

The next example will deal with detection of hardly visible WC-WC boundaries in a WC-Co cermet. None of the algorithms discussed until now gives satisfactory detection in this case. So, we will apply a completely new algorithm:

- The Co phase is very easily detected, as it is black (Fig. 4.16a) and can be succesfully detected using a simple threshold. The WC grains exhibit approximately homogeneous gray levels and only at their grain boundaries is a narrow layer of pixels with a more significant difference in gray levels observed
- for detection of this layer we have developed a user-defined filter, sensitive to the gradient in the gray levels. In order to cover all the directions, the gradient is checked at four orientations: 0°, 45°, 90° and 135°, denoted as grad0, grad45, grad90 and grad135, respectively. For detection of grad0 (Fig. 4.16b) and grad90 (Fig. 4.16c), the following two filter matrices of size 2x3 are used:

1	1	1
-1	-1	-1

and

-1	1
-1	1
-1	1

- for detection of the grad45 and grad135 gradients (not shown in Fig. 4.16), another set of two new matrices of size 4x4 is applied:

0	1	0	0
-1	0	1	0
0	-1	0	1
0	0	-1	0

and

0	0	1	0
0	1	0	-1
1	0	-1	0
0	-1	0	0

- the gradient images give both positive and negative gradient values, depending on the local gradient orientation. As we are interested in detecting only the existence of any gradient, irrespective of its orientation or magnitude, we compute the absolute value of the partial gradient images, giving information about four different orientations within the image. These temporary images are not shown in Fig. 4.16

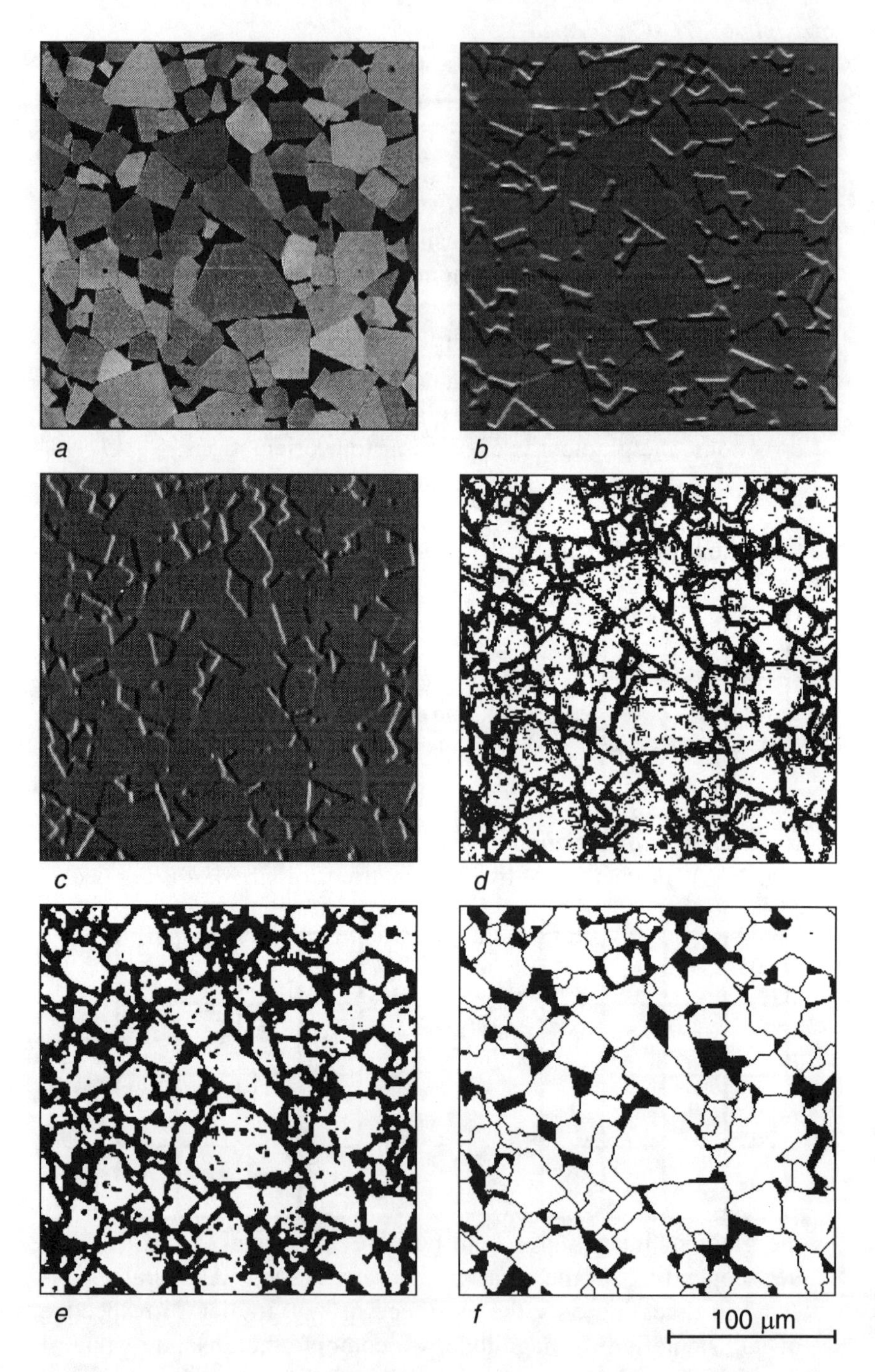

Fig. 4.16. Detection of boundaries between the WC grains in a WC-Co cermet. Initial image (a), grad0 (b), grad90 (c), overall gradient after binarization (d), median filter (e) and final detection after SKIZ (f).

- the final gradient image is constructed as a maximum of all the four intermediate gradient images, obtained after computing the abstract values. In arithmetic functions we will find the appropriate one, returning a maximum of two input images. Using it for creation of a maximum of four images should not be difficult. The final gradient image can be binarized to produce an image with roughly detected grain boundaries (Fig. 4.16d)
- as in previous examples, we can filter out the noise using the median filter. The filtered image (Fig. 4.16e) can be used for a subsequent SKIZ in order to get the finally detected grains
- The final image is produced from two images: pores detected by binarization (not shown here) and grains obtained using SKIZ. The final image (Fig. 4.16f) can be produced by a logical AND (assuming that pores are black in the binary image).

The final result is quite satisfactory, however some grain boundaries are still not detected. It can be slightly improved after the use of watershed segmentation to the image shown in Fig. 4.16e. This is a solution similar to that applied in the previous example. Therefore, it is not discussed in this example.

Example 7: Grains in a high-speed steel.

In this example we will use various solutions already described in the previous examples, but for the first time we will use as a data source two entirely different images, taken from the same area. As in previous examples, we will analyze step-by-step the algorithm that is illustrated in the figures showing intermediate steps of the algorithm:

- we will start from the initial image, taken by SEM (Fig. 4.17a). This image has a very poor contrast and, therefore, any detection of grain boundaries is very difficult
- we can try to improve the contrast using various filters, but we should remember that the majority of filters simultaneously improve contrast and amplify the noise present in the image. Here (Fig. 4.17b) the contrast is improved thanks to the FFT transformation and filtering out the lowest frequencies
- the grain boundaries, in spite of the FFT filtering, cannot be detected by binarization and, therefore, we will use the gradient procedure similar to that presented in the previous example. Here we can observe two partial gradient images: grad45 (Fig. 4.17c) and grad90 (Fig. 4.17d). The final gradient image, being a sum of partial gradients, is shown in Fig. 4.17e

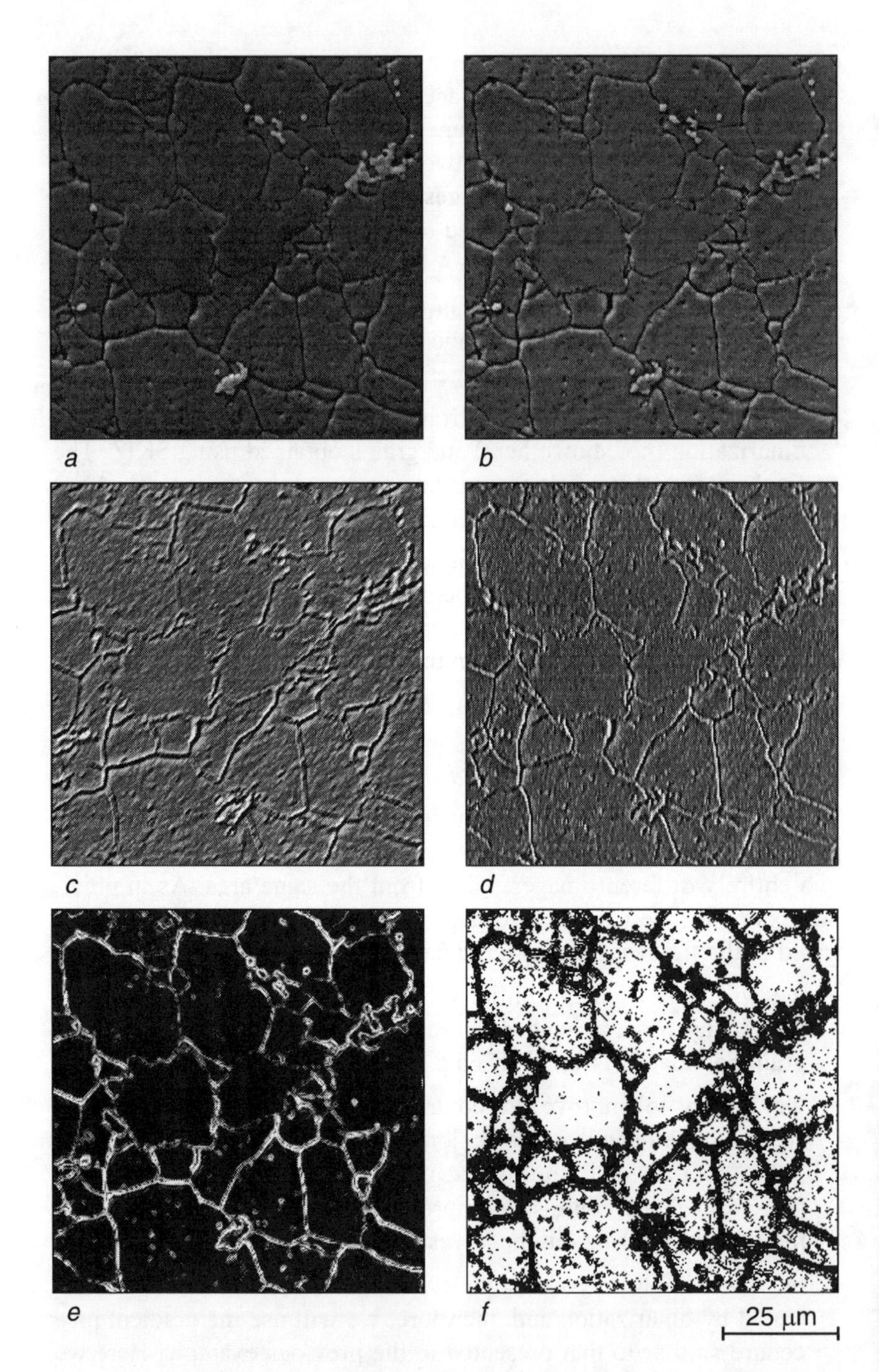

Fig. 4.17. Preliminary steps in analysis of a high-speed steel: initial SEM image (a), enhanced FFT filtered image (b), partial gradients (c, d), the final gradient image (e) and binarized gradient image (f).

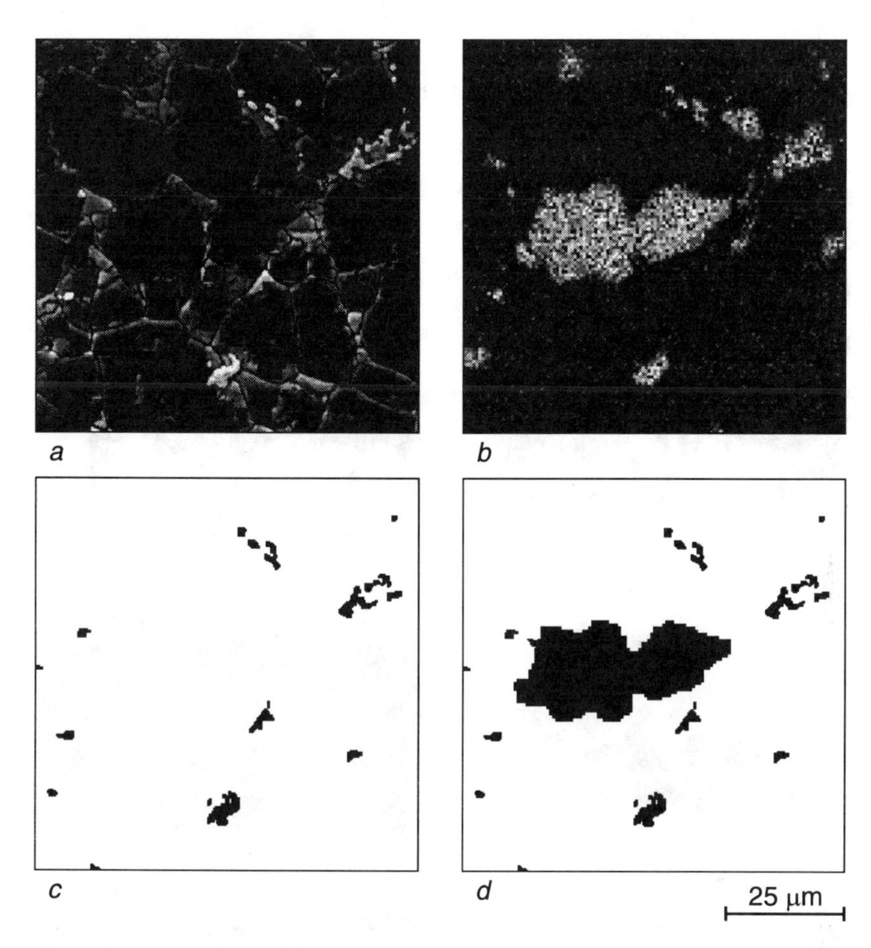

Fig. 4.18. Detection of carbides: enhanced, small, bright carbides (a), additional image showing the surface distribution of Mo (b), detected small carbides (c) and all the carbides detected (d).

- thresholding of the gradient image (Fig. 4.17f) give us the starting point for further analysis
- we will now switch for a moment to detection of the carbides. We can easily enhance the small carbides by subtracting the eroded initial image from just the initial image. It is, obviously, the first step of a white top-hat transformation (Fig. 4.18a). The subsequent binarization gives the small carbides, as shown in Fig. 4.18c
- we will have a little difficulty during detection of the largest carbides. Fortunately, we have at our disposal the surface map of molybdenum content (Fig. 4.18b). Thresholding and subsequent closing (it causes hole filling) followed by opening (removal of small particles) gives correctly detected large carbides

- logical summation (OR) of the images of small and large carbides allows us to get the final detection of all the carbides (Fig. 4.18d)

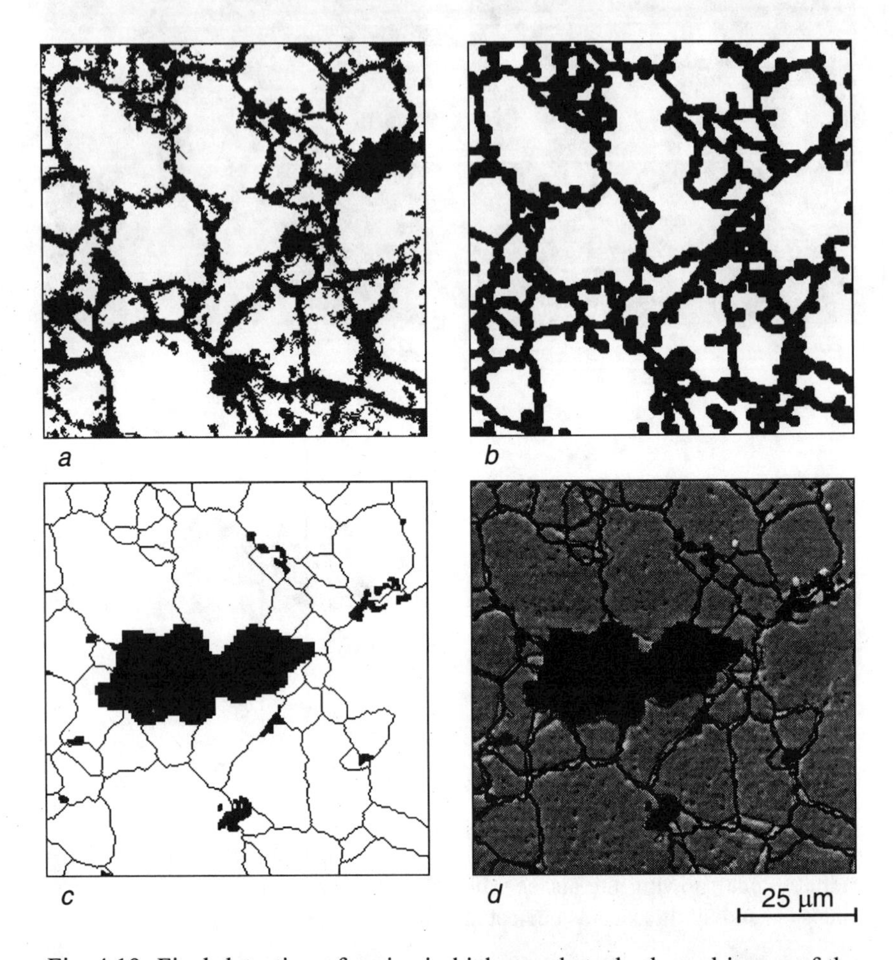

Fig. 4.19. Final detection of grains in high-speed steel: cleaned image of the binarized gradient (a), SKIZ followed by erosion (b), final detection (c) and the final detection overlaid over the initial image (d).

- the binarized gradient image (Fig. 4.17f) looks very noisy. Fortunately, it contains a continuous network of roughly detected grain boundaries and, therefore, we can clean up this image by simple hole filling (Fig. 4.19a)
- the grains can be built now using the SKIZ transformation. There are still some very small features in the region of the expected grain boundaries. If we remove them by erosion, we will probably affect the shape of the grains detected. If we do not remove these small elements, we will get some very small extra grains. In order

to remove them we will apply some erosion to the image after a SKIZ (Fig. 4.19b) and repeat the SKIZ operation to get the final image of grains

- combining the images of grains and carbides (Fig. 4.18d) by logical summation (OR) will produce the final detection of grains together with the carbides (Fig. 4.19c)
- the final detection seems to be a good response to the initial image. In order to check the quality of the whole, quite long, algorithm the detected features are overlaid over the initial image (Fig. 4.19d).

Example 8: A tricky solution to a nearly hopeless case

This last example will deal with grain boundary detection in a compound ceramic ZrO_2 with 20% of WC. Due to its mechanical and chemical resistance, this material is extremely difficult for specimen preparation and the images obtained are not very clear (see Fig. 4.20a). Binarization can give us the correct detection of pores (Fig. 4.20b), but a subsequent watershed detection can give somewhat artificial effects.

If we have such a poor quality image, we face the temptation to apply any sharpening procedure. It leads to results that look really hopeless (see Fig. 4.20c). To make things worse, it is often the only image at our disposal and our colleagues ask us (you are our last chance, etc.) to do something with this problem. Quite probably you will try to apply binarization to this case. It is natural to start with the simplest operations and additionally this binary image gives us some information on future difficulties in correct detection (see Fig. 4.20d).

The situation we have in the image in Fig. 4.20d is really frustrating. We see clearly the regions of grain boundaries but any attempt to detect the continuous network of grains fails. If we apply erosion in order to remove some points in the center of the image, we will get results worse than those shown in Fig. 4.20b. Applying a hole filling procedure to the negative of this image will produce a uniformly filled region, because we have no continuous network of pixels. We can try to apply the median filtering as in some of the previous examples but, due to the high density of white pixels it also will not bring us success. So, it is a really interesting case and we will analyze a nice solution to this problem. We will take Fig. 4.20d as the initial image for our analysis.

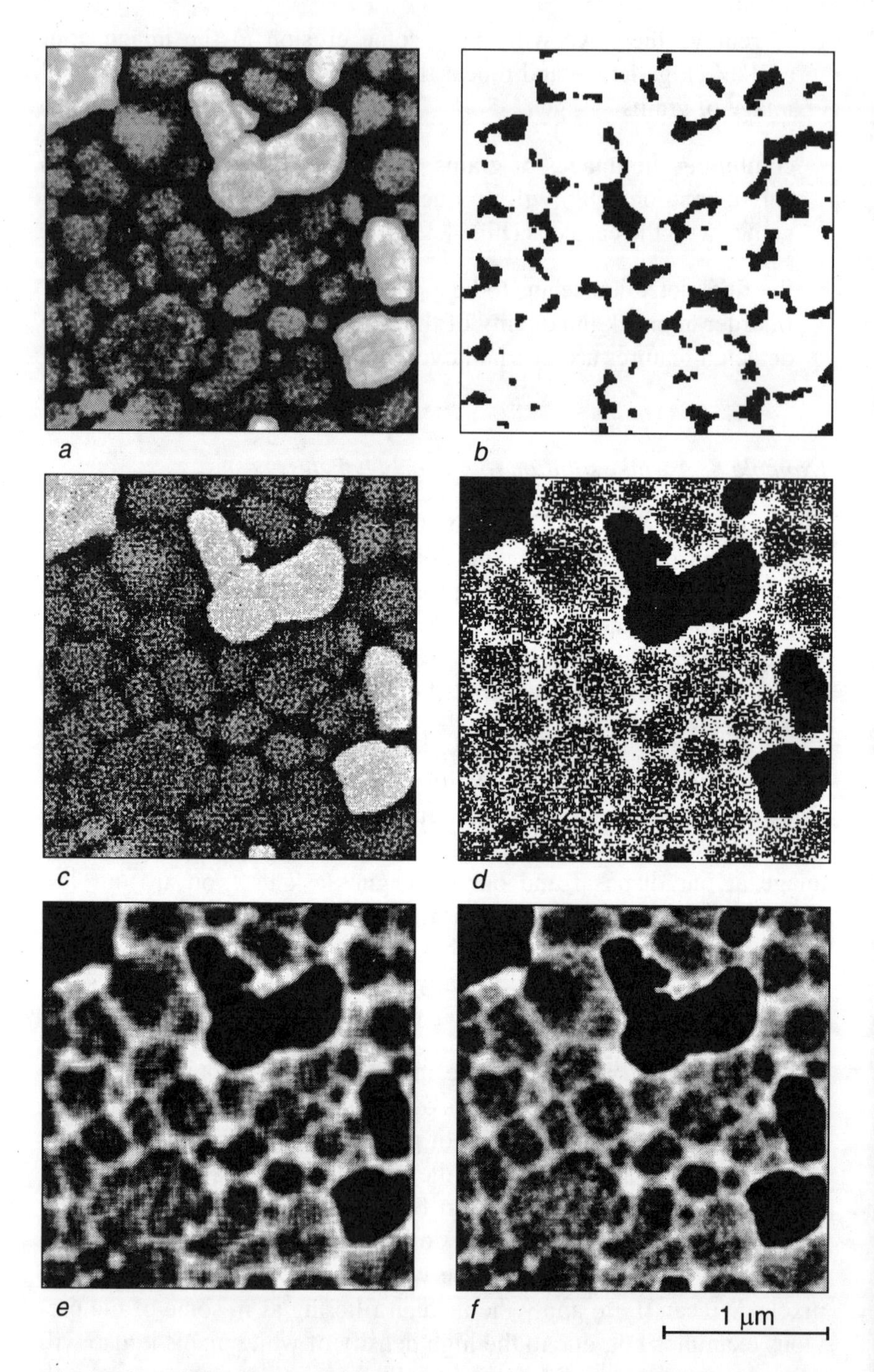

Fig. 4.20. A difficult example of the microstructure of a ceramic (a), detected pores (b), sharpened image (c), its binarized version used as initial image for further analysis (d) and density maps: square (e) and octagonal (f).

If you look for the solution to this problem, you probably will note that your vision system recognizes the locations of possible grain boundaries in Fig. 4.20d because at these regions we have a higher density of white pixels. So, we will try to develop a suitable filter that will be sensitive to the pixel density. The simplest solution seems to be the mean number of white pixels in an area of large enough size, for example, 7x7 pixels (a smaller test area will be much too sensitive to any local irregularities). You can use for this purpose the filter returning the arithmetic mean or some low-pass filters (but check their matrices!). You can also apply a user-defined filter, filling the whole 7x7 matrix with 1 (one). As a result, you will get an approximate *pixel density map*, as shown in Fig. 4.20e. Unfortunately, this density map still exhibits too much noise.

We are mainly interested in detecting the regions of highest density, representing the network of grain boundaries. If we imagine a circle surrounding any point on the grain boundary, this boundary should cross this circle very close to the diameter. Therefore, the corner pixels in the 7x7 matrix used previously can be thrown away. So, for our purposes we will use an octagonal filter, with a matrix defined as shown below:

0	0	0	1	0	0	0
0	1	1	1	1	1	0
0	1	1	1	1	1	0
1	1	1	1	1	1	1
0	1	1	1	1	1	0
0	1	1	1	1	1	0
0	0	0	1	0	0	0

Application of this octagonal matrix gives a significantly smoother pixel density map (see Figs. 4.20e and f), however, it is still too jagged. In order to improve the map we can apply the same filter once more or even twice. The pixel density map after three cycles of computation with the octagonal matrix is shown in Fig. 4.22a. Three profiles, shown in Fig. 4.21, allow us to trace the density map development in a more quantitative way.

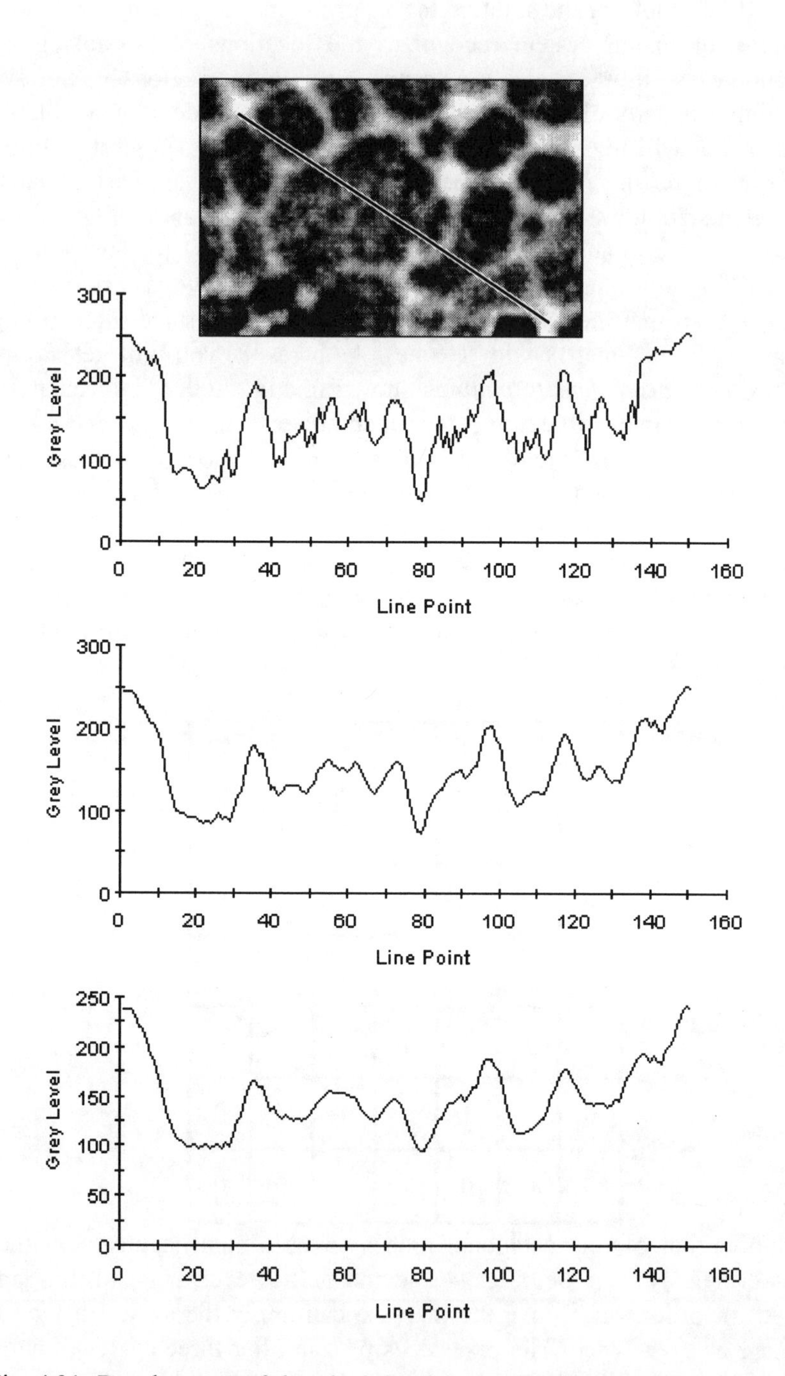

Fig. 4.21. Development of the pixel density map. Profiles computed along a line shown in the top image. Profiles represent the following stages of computation: first (top), second (middle) and third (bottom).

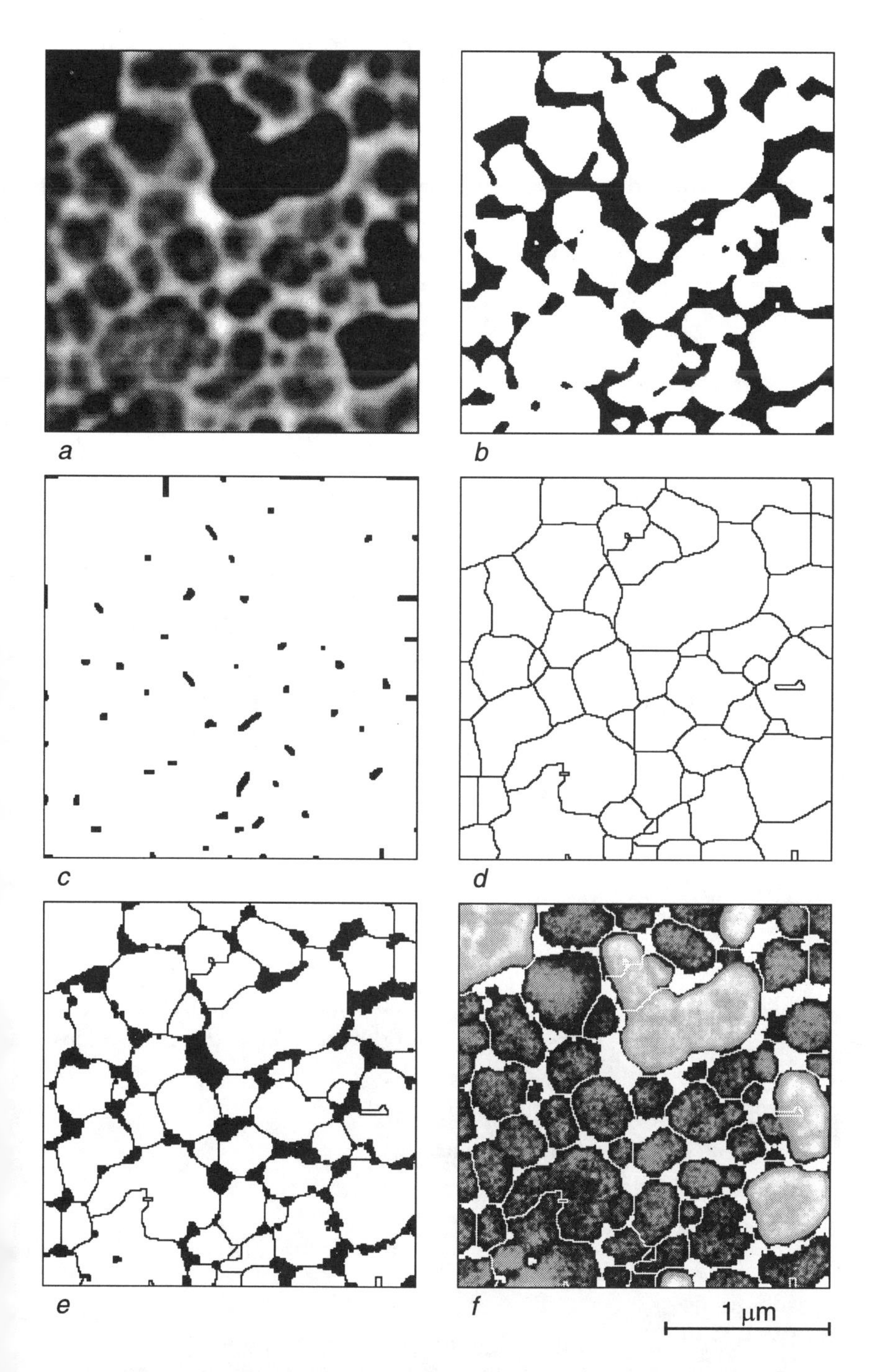

Fig. 4.22. Detection of grains - continued from Fig. 4.20. Initial pixel density map (a), detected grains (b), ultimate erosion (c), watershed segmentation (d), final result (e) overlaid over initial image (f).

The pixel density map after three cycles of convoluting with the octagonal matrix is now smooth enough to enable the final detection of grain boundaries:

- the best method of detection seems to be an application of the watershed. In order to avoid unnecessary over-segmentation we will apply the constrained watershed and for this purpose we need appropriate markers. The markers are prepared in two steps. The first step is binarization with the threshold large enough to detect traces of all the grains. It leads to grain prototypes glued together (Fig. 4.22b). Fortunately, the pixel density map is so smooth that the detected grains are convex and easy to mark with help from an ultimate erosion (Fig. 4.22c)
- the next step in our analysis will be the construction of grains using a constrained watershed detection. We will use the pixel density map (Fig. 4.22a) as a distance image. The resulting image of grains is shown in Fig. 4.22d. To produce the finally detected grains we will use the image of pores (Fig. 4.20b) and the logical AND function. The final detection is presented in Fig. 4.22e
- we can easily check the quality of the final detection by looking at the detected image overlying the initial image. For better contrast, the detected image is plotted in white (Fig. 4.22f). The final detection should be rated as very good, especially if we take into consideration the extremely low quality of the initial image.

4.2 Other features detected like grains

The contents of Section 4.1 show a large variety of possible algorithms used for grain boundary detection. Exactly the same techniques can be applied for detection of other features in a material microstructure, for example:

- dimples on ductile fracture surfaces
- facets on brittle fracture surfaces
- isolated particles, fibers or pores
- colonies of similarly oriented features
- networks of precipitates or cracks
- sub-grains or
- layers in coated or highly inhomogeneous structures.

The above list can be extensively updated. Obviously, it is impossible to cover all the existing structures, but we tried in this chapter to show numerous and universal solutions.

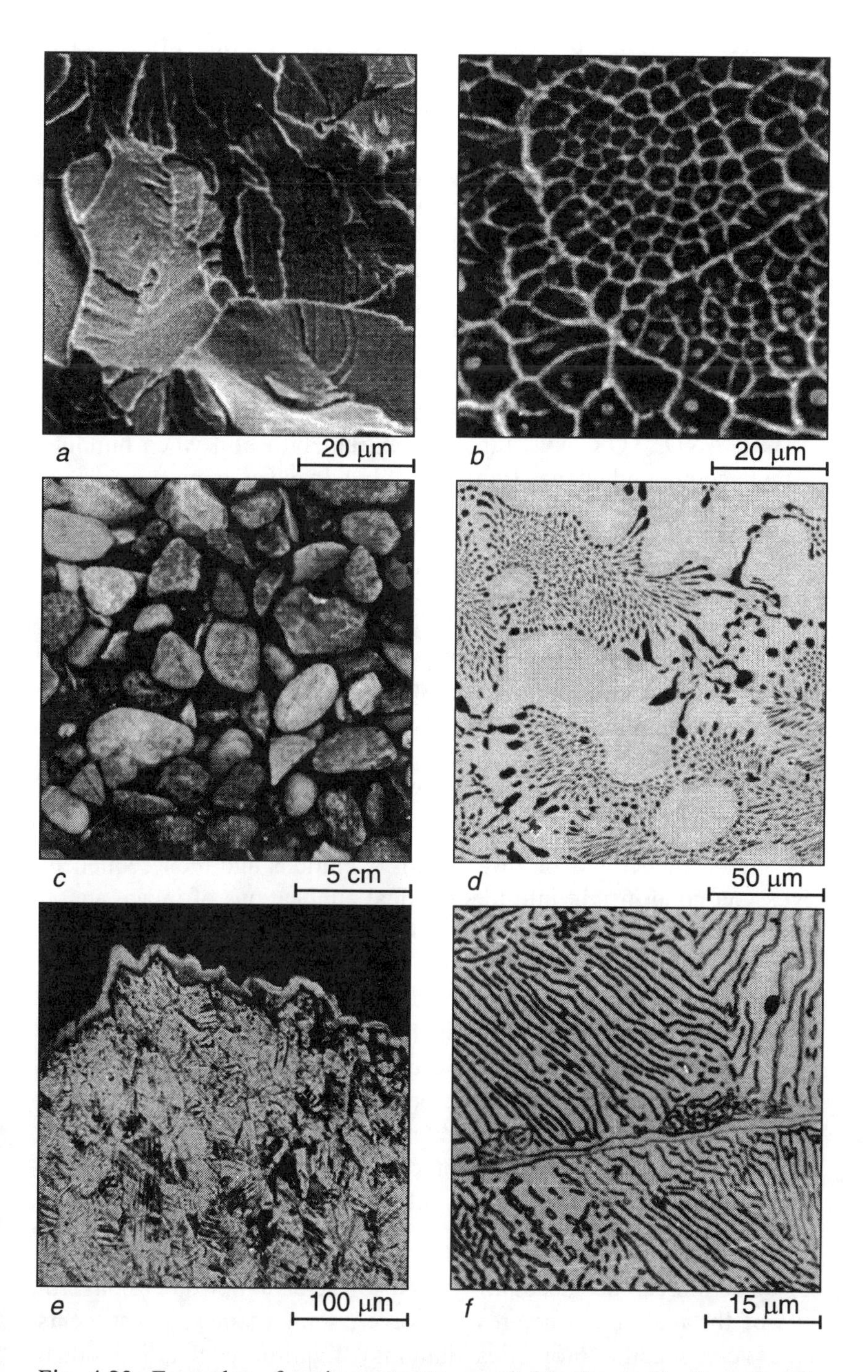

Fig. 4.23. Examples of various structures suitable for application of grain boundary detection methods: ductile (a) and brittle (b) fracture surfaces, gravel particles (c), eutectic (d), fracture profile (e) and lamellar structure (f).

Seven examples presented in the previous section offer an extended range of detection algorithms that can be adapted to many other structures. In these examples we have used, in various combinations, a wide spectrum of image transformations. In fact, the list of procedures applied covers all the main groups of operations met in image analysis, in particular:

- binarization
- filters for sharpening, blurring, noise reduction or edge detection, as well as user-defined filters
- arithmetic (add, subtract, maximum) and logical transformations (OR, AND, XOR, NOT) that are extremely useful when building new, compound images from a few already existing ones
- Fourier transformations together with specific filters used for modifying the Fourier spectrum prior to the inverse Fourier transformation
- morphological transformations: erosion and dilation, ultimate erosion, top-hat, hole filling, reconstruction, skeletonization or watershed detection, that are often decisive for the final success of the whole set of operations and
- complex algorithms, for example, for shade correction.

Such a rich collection of different operations and their sequences gives you good insight into the practical applications of image analysis techniques. Probably you have noted from these cases that successful application of image analysis tools requires a completely new way of treating material structures. We cannot think any longer about pores, fibers, grains or eutectics as these terms are not understandable to a computer. Instead, we should think about gray values, textures, densities and gradients which can be recognized by the machine. Our success in image analysis depends mainly on the ability and skill of translating the structural problems of materials into logical and geometrical problems of the sets of pixels - computer images.

As was already mentioned, the solutions applied to grain boundary detection can be successfully used for detection of other constituents of the material structures. However, some of these constituents have very specific properties that may require special, individual treatment. The next sections of this chapter will be devoted to the analysis of just such cases.

4.3 *Pores and isolated particles*

Detection of pores and isolated particles usually does not yield any greater difficulties. In most cases, pores, due to the physical properties of microscopes, are the darkest regions of the image and can be easily detected by simple binarization (Fig. 4.24). The main problem in detection of pores, arising during examination of the polished sections, is their correct preparation prior to microscopic examination. In the case of very soft materials the edges can be plastically deformed, leading to visual compression or even closing of the pores. By contrast, in the case of very hard and brittle materials, the abrasive particles can crush and remove a part of the material located at the edge of the pore, leading to visual enlargement of the pores (Fig. 4.25). Usually, image analysis methods do not allow one to decide if the pores are correctly prepared from the viewpoint of metallography.

Fig. 4.24. An example of a polished surface after plastic deformation. Visible black pores around the particles.

If we predict the necessity of analysis of isolated particles, the best solution is to use specimens adequately prepared for this unique analysis. In the most convenient case, a polished section without any etching can be used for this purpose. The most common examples are non-metallic inclusions in steels, graphite precipitates in cast iron (Fig. 4.26, see Fig. 3.12), gravel in concrete or fiber sections in numerous fiber-reinforced composites. In these fortunate cases the particles can be detected after simple binarization. The only restriction is the absolute obligation of the highest possible quality of the specimens. Any errors in specimen preparation forces subsequent cleaning of the binary image that can lead to errors in the analysis. The main problem during binarization is the choice of the threshold level. Some guidelines on this point were presented in Section 2.4.

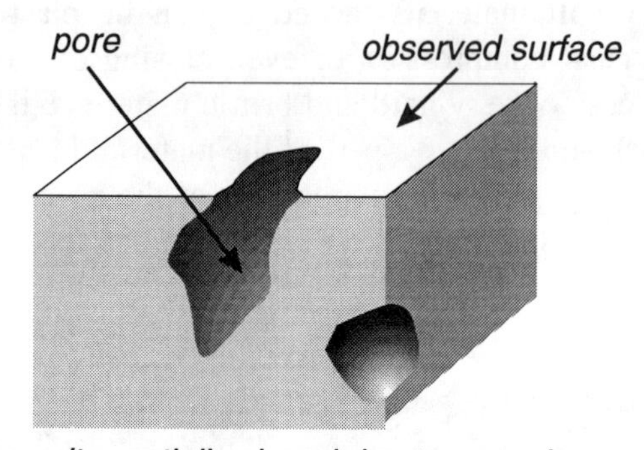

porosity partially closed due to smearing

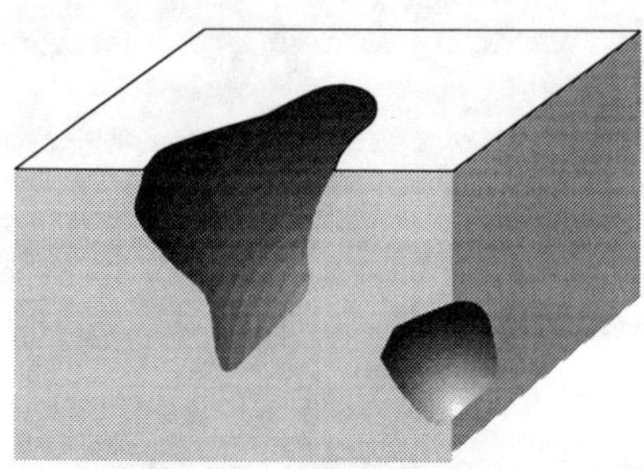

porosity not distorted

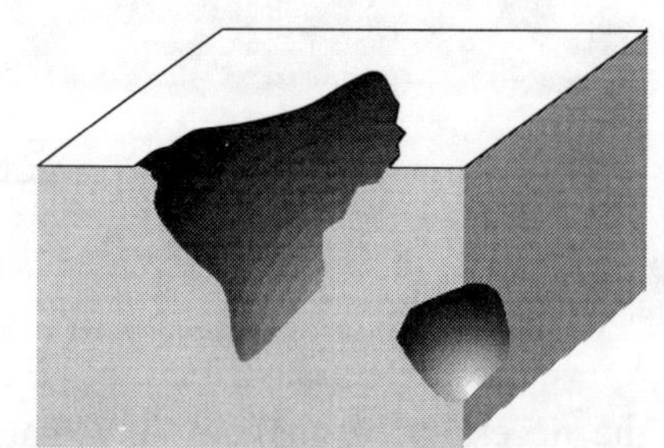

porosity enlarged due to brittle fracture

Fig. 4.25. Effect of specimen preparation on the pores (schematically). Smearing closes porosity in soft materials (a), specimen optimally prepared for observation (b) and pores enlarged due to decohesion of a very brittle material.

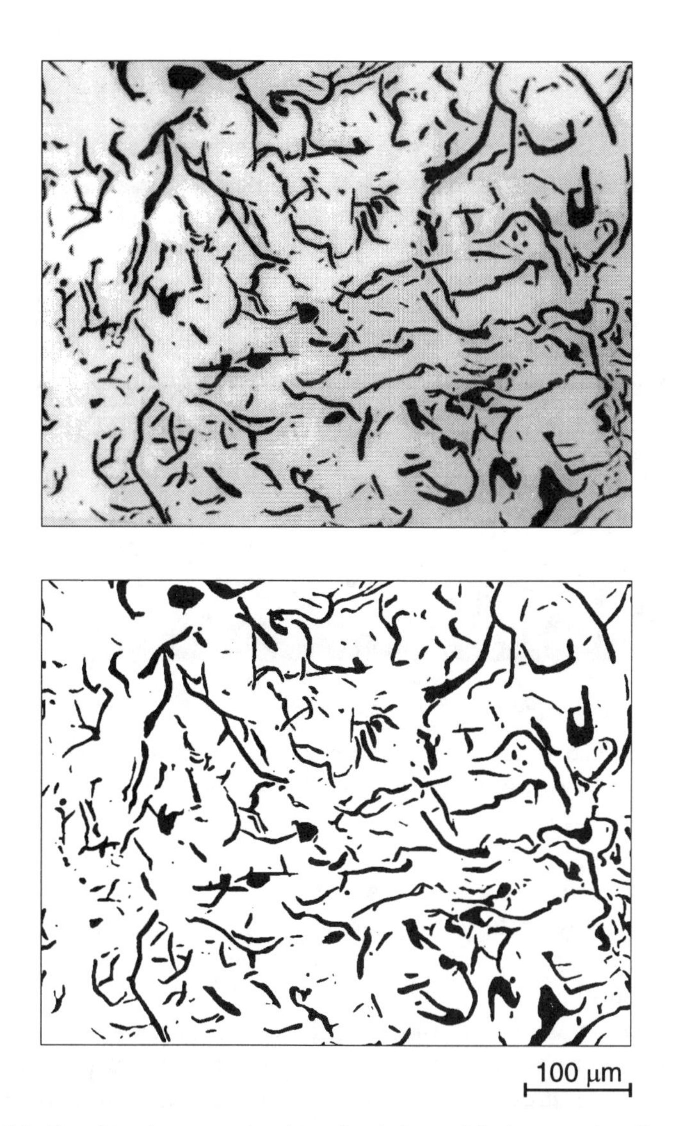

Fig. 4.26. Graphite in gray cast iron (top) is an ideal example of particles which are easy to detect by binarization (bottom).

Another frequent case is from specimens that require special treatment prior to any analysis. In most cases, this will be selective etching which enhances the contrast from the phases under consideration. A good example is shown in Fig. 4.27. Ordinary etching of this duplex cast steel will differentiate only the austenite and ferrite, leaving carbides completely invisible. Application of the Beraha solvent allows us a clear distinction between the phases. This example indicates once more how important the proper specimen preparation is. In fact, the better prepared the specimen is, the less image treatment is necessary and the more precise the results of the whole analysis.

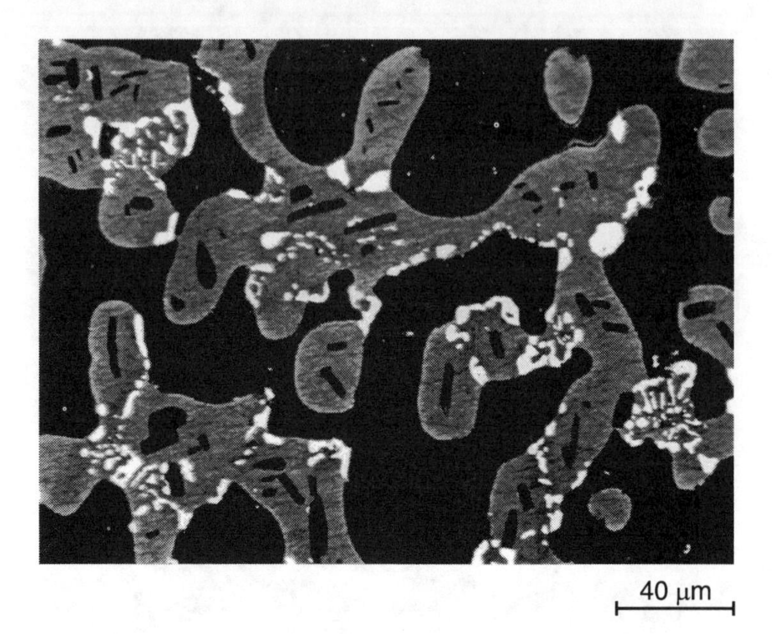

Fig. 4.27. Selective etching enables easy detection of carbides in a austenitic-ferritic duplex cast steel. Austenite (dark gray), ferrite (light gray) and carbides (white).

Similar effects can be observed in the case of electron microscopy. Especially useful is the scanning electron microscope, as we can combine various signals or even various images, including the surface maps illustrating the distribution of elements. Very often local fluctuations in the contrast of the image produced by the specimen microrelief lead to local irregularities in the gray levels that disable detection by simple thresholding (Fig. 4.28). Fortunately, this problem can usually be successfully solved by application of a top-hat transformation.

In transmission electron microscopy there are also available various tools for image formation that can help visualization of the features under investigation. However, in some cases, when subtle phenomena are investigated, even the most advanced techniques produce images of relatively poor contrast. An illustrative example is shown in Fig. 4.29, in which we can observe a helium-implanted sample of Nimonic PE16 at high magnification. The size distribution of helium bubbles needs to be investigated. Proper detection enables quantitative analysis of the content and size distribution of the bubbles.

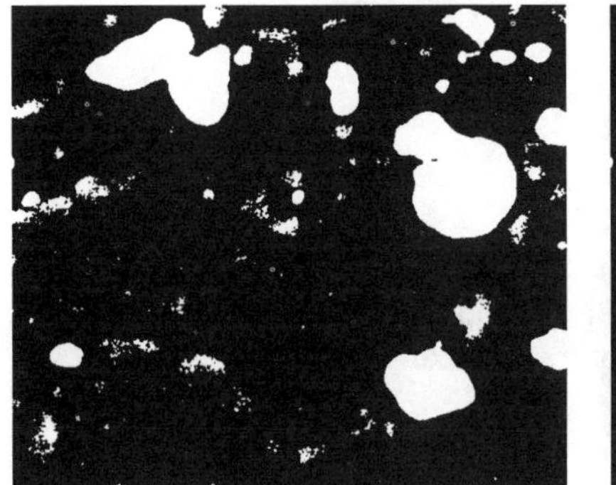

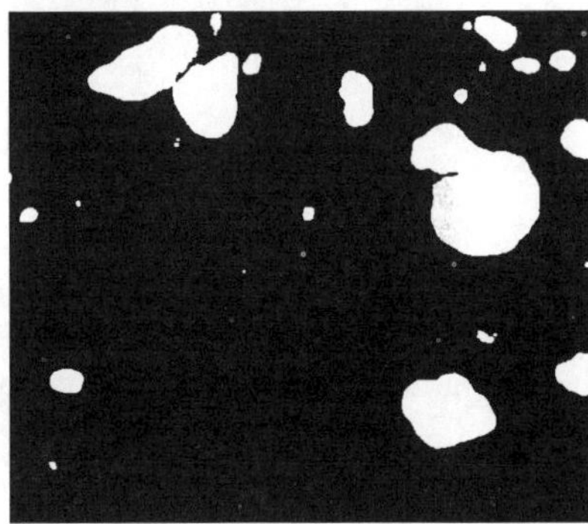

Fig. 4.28. A ZrO_2-WC ceramic (top). Detection of the WC precipitates by thresholding is not very satisfactory (bottom left) but the top-hat transformation leads to correct detection (bottom right).

Certainly, detection of isolated particles is not always as simple as in the examples presented. For example, correct detection of the inclusions shown in this section in Fig. 4.24 would require a treatment similar to that presented for grain boundary detection. So, any rigid classification of structures is impossible, because we often meet ones that are a very complex mixture of simplified, model cases. The remarks presented in this section answer the most typical cases when the main goal of our analysis is just particle recognition and the whole specimen preparation process is oriented towards visualization of these particles. Many people, having little knowledge of image analysis think that contemporary tools can pull-out the desired information from any image, even if it is not correctly prepared. This is, obviously, not possible.

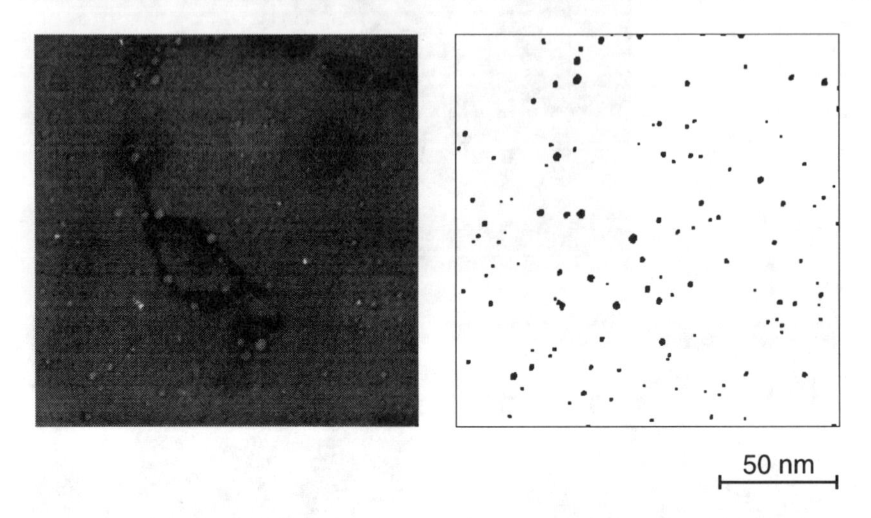

Fig. 4.29. The structure of Nimonic PE16, implanted by helium (left) and the helium bubbles detected by the top-hat transformation.

4.4. Chains and colonies

When analyzing the inhomogeneity of properties we often want to check the existence of clusters or colonies of particles or other structural constituents. Simultaneously, when discussing the anisotropy of plastically deformed materials we frequently want to detect any chains of particles oriented along the direction of plastic deformation.

Any analysis of colonies or chains is extremely difficult, if not impossible, in gray-scale images. The most natural way seems to be the use of binary images with finally detected particles. Detection of the features discussed in this section is in such circumstances much easier. For a final success we need only to define precisely enough the mathematical or geometrical criteria necessary to recognize a given set of particles as a cluster or chain.

From the above remarks we can state that the final effect of the analysis depends in this case on our knowledge of material engineering rather than on image analysis. However, some intuition can be very helpful, because the decisive factors are not always directly visible. In this section we will show a few examples of the evaluation of chains and colonies starting from the binary image. The procedures of particle detection leading to these binary images can be formed on the basis of other parts of this chapter, as well as the previous ones.

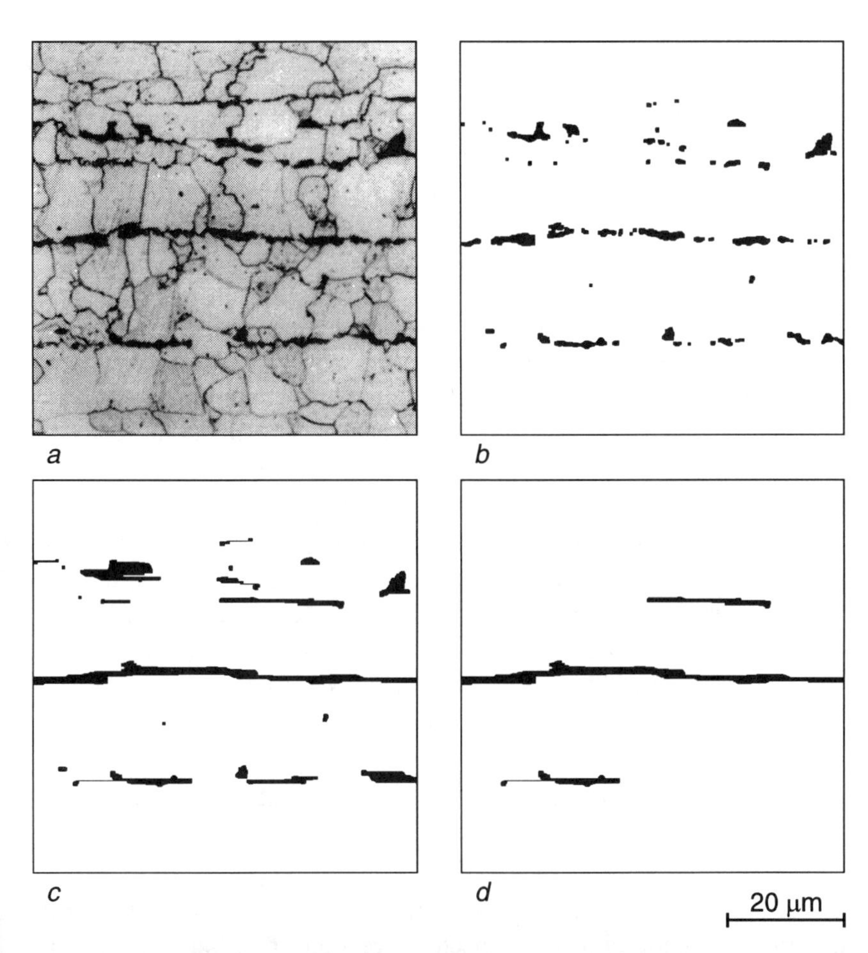

Fig. 4.30. Non-metallic inclusions in low carbon steel. The microstructure (a), binary image of inclusions (b), detected chains (c) and large chains (d).

Chains of particles are usually formed during plastic deformation introduced, for example, by rolling. They significantly affect the toughness and, even more, the anisotropy of the mechanical properties of the material. An example of non-metallic inclusions in a low carbon steel arranged in chains is shown in Fig. 4.30. To show the possible interaction between the grains of ferrite and inclusions the polished section was etched (Fig. 4.30a) and the inclusions were detected using a top-hat transformation (Fig. 4.30b). Usually, the contrast between inclusions and the polished matrix is very good. Therefore, in quality control tests only the un-etched specimens are used. This minimizes the digitization errors, as etching always deteriorates the matrix-inclusion interface.

Detection of chains is very simple. Application of the closing procedure will join any particles that are placed close to each other and produce the image of chains (Fig. 4.30c). Further filtering can lead, for example, to detection of the largest chains (Fig. 4.30d).

A similar approach is possible in the case of carbides in tool steel (Fig. 4.31). Carbides are responsible for the cutting properties of the tool. Unfortunately, they are simultaneously hard and very brittle. Therefore, in order to ensure the optimum properties, carbides should be uniformly distributed in the material volume and surrounded by relatively soft and plastic matrices. This last property ensures sufficient toughness of the whole material. If the carbide precipitates are distributed in chains (usually called bands), the material tends to fracture along them and the lifetime of any tools made of this material dramatically decreases.

Due to the high volume fraction of carbides, analysis of all of the visible precipitates will give no effect (Fig. 4.31b). Thus, we usually analyze only the largest carbides. They can be easily distinguished by opening or, if we want to preserve the original shape of precipitates, by erosion followed by reconstruction (Fig. 4.31c).

Now we are in a position to try to detect the bands of carbides. Application of the closing procedure was a good solution in the case of elongated inclusions (see Fig. 4.30). In the case of carbides we do not observe any elongation of the particles and closing leads to somewhat unnatural results (Fig. 4.31d).

Some improvement can be observed if we apply *linear closing* in the direction parallel to the expected direction of carbide bands (Fig. 4.31e). Linear closing (also erosion, dilation and opening) is very similar to the classical transformations described in Section 2.5. The only difference is in the shape of the structuring element that is, in the case of linear transformations, a line segment. If we compare the results of linear closing parallel (Fig. 4.31e) and perpendicular (Fig. 4.31f) to the direction of bands, we will note that these linear transformations are useful for the detection of any linearly oriented features. Nevertheless, the result of the detection procedure is still rather poor.

An acceptable solution can be achieved, as usual, by combining many different operations. This is presented (on another piece of material similar to that shown in Fig. 4.31) in Fig. 4.32.

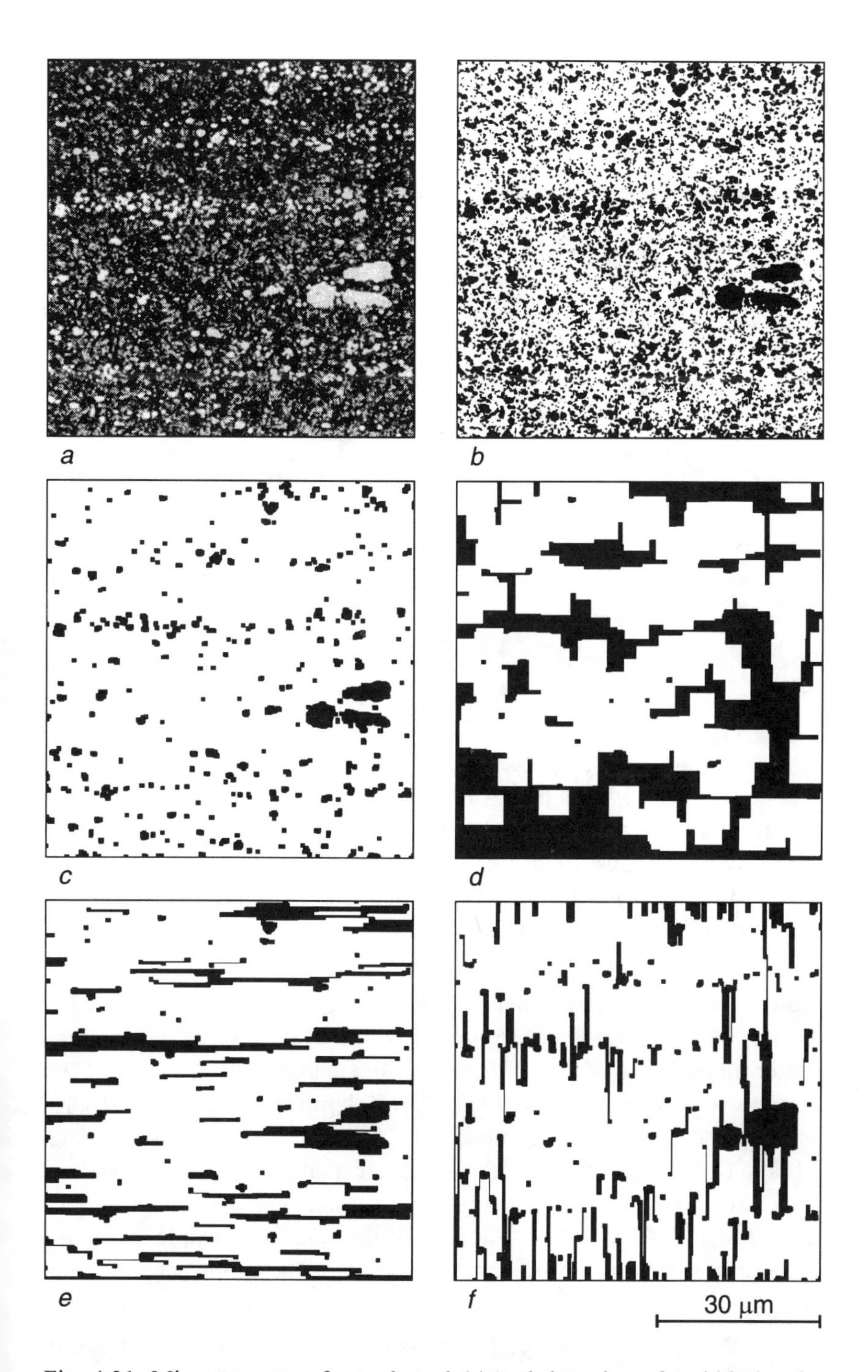

Fig. 4.31. Microstructure of a tool steel (a) and detection of carbide bands: detected carbides (b), the largest carbides (c), bands created from large carbides by closing (d) and horizontal (e) or vertical (f) linear closing.

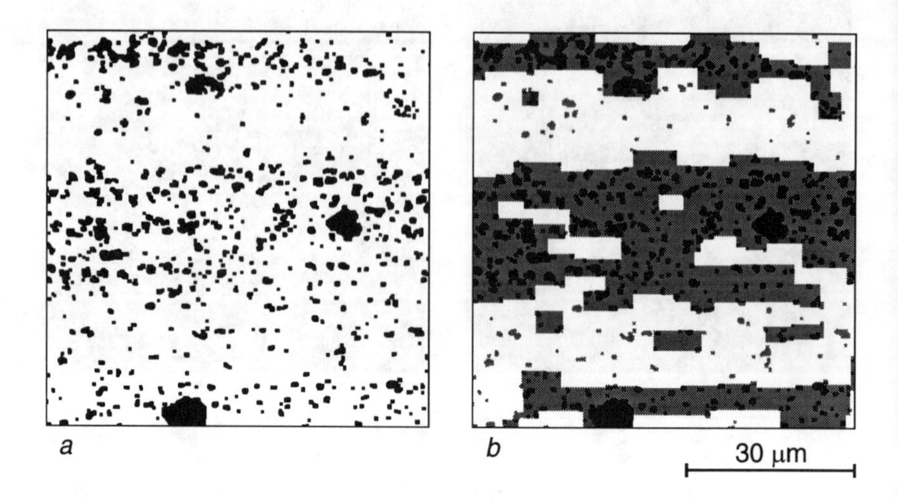

Fig. 4.32. Detected carbides (a) and the same carbides superimposed on the detected bands, drawn in gray (b).

Now we will analyze in more detail, the detection of carbide bands in Fig. 4.32. First, we will check the effect of closing on the detected regions. We apply the following closings of different sizes: 5 (Fig. 4.33a), 10 - this means two times greater (Fig. 4.33b) and 20 - this means four times greater (Fig. 4.33c). It is clearly visible that we cannot get good results in this way. With the increasing closing size we get more and more continuous bands but simultaneously we build unnecessary bridges connecting parallel bands.

Linear closing (Fig. 4.33d) gives quite promising results, but the bands obtained are still discontinuous and not suitable for further analysis. By adding a small (here of size 10) closing we get a much better result (Fig. 4.33e). However, the bands still appear a little strange, with artificially looking narrow branches. We can easily remove these branches by a subsequent (here also of size 10) opening. This set of simple morphological transformations - there are in fact only erosions and dilations - gives the final image of the bands of carbides (Fig. 4.33f). Some additional filtering, for example, a median of a large size can lead to smoother edges of the bands. Significantly smoother detection can also be obtained using a hexagonal grid.

Looking at the detected band overlying the carbides (Fig. 4.32b) you will probably note that the result is quite good. Moreover, in contrary to the manual detection or drawing of the bands, it is objective and reproducible.

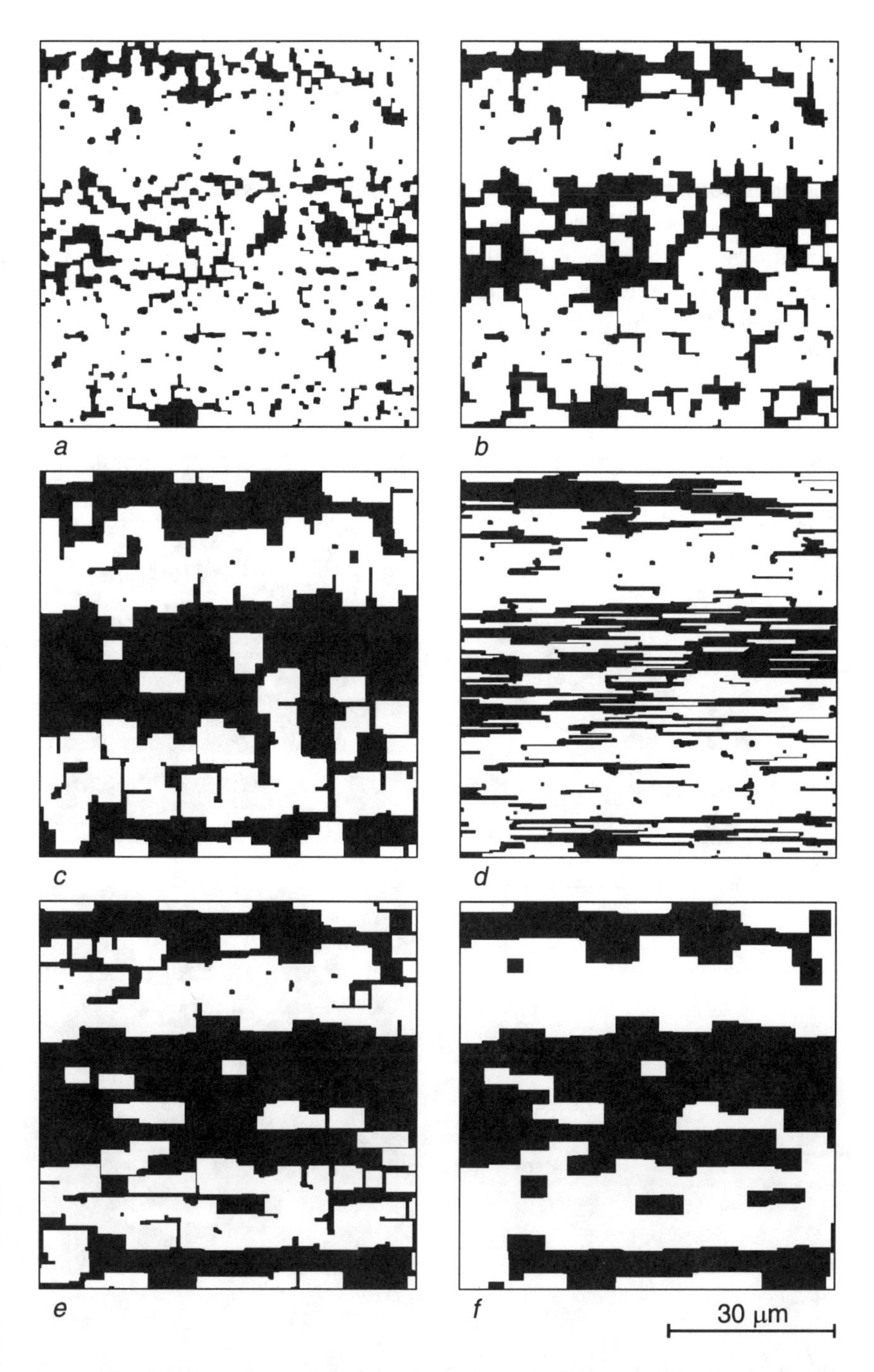

Fig. 4.33. Various stages of detection of the carbide bands. Closings of various sizes (a, b, c) and linear closing (d) followed by closing (e) and opening (f) of small size.

Ordinary closing is still a very good tool for the detection of clustered equiaxial particles like, for example, graphite nodules in a ferritic ductile cast iron (Fig. 4.34a). Closing will form a set of clusters, as shown in Fig. 4.34b. The clusters are very close to each other and therefore plotted in different gray tones for easier recognition of the separated sets. Worth noticing is the fact that the clusters shown in this image are relatively small in comparison with the graphite. Any attempt to produce larger clusters will lead to including almost all the image pixels in the clusters. This can be interpreted as a uniform distribution of the graphite nodules.

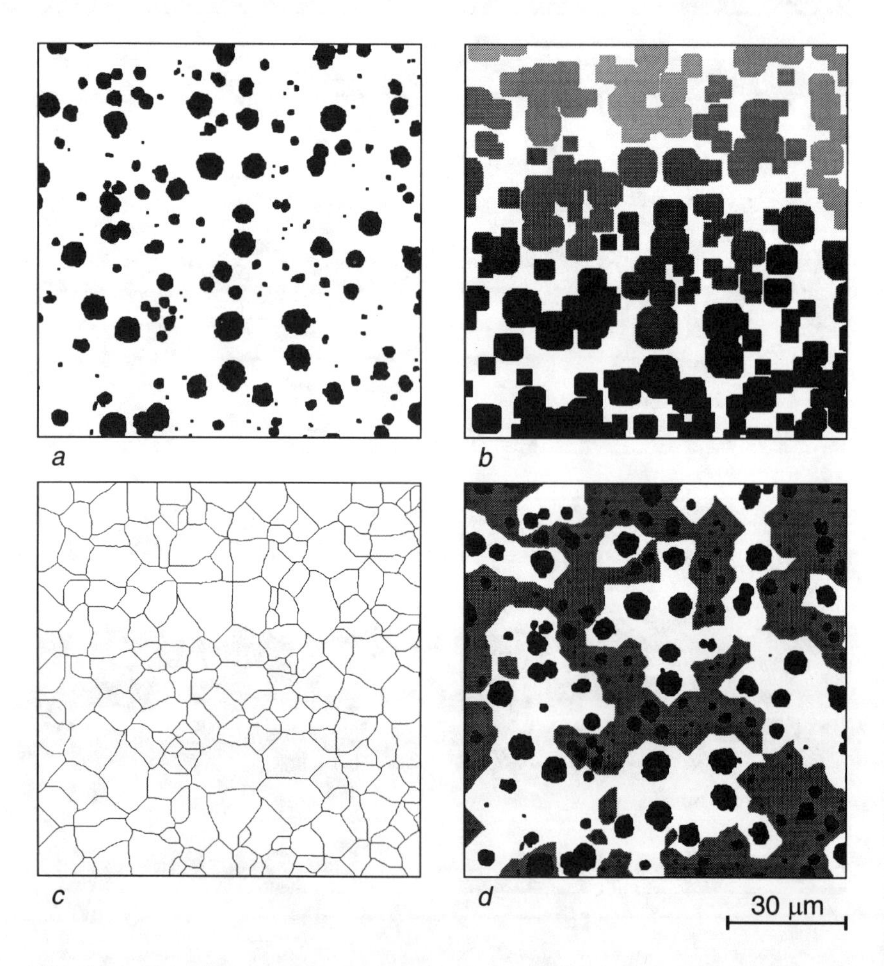

Fig. 4.34. Two different approaches for cluster detection of graphite nodules. Graphite phase (a), clusters, plotted in different gray levels, formed by a small closing (b), influence zones of the graphite nodules (c) and another set of clusters, formed from filtered zones of influence (d).

There is also possible an entirely different approach for the detection of the graphite clusters. Using the SKIZ operation we can build the zones of influence for all the graphite nodules (Fig. 4.34c). The zone of influence for any graphite nodule can be interpreted as a set of points that are closer to this nodule than to any other one. On the basis of the influence zones we can create another set of clusters, shown in Fig. 4.34d. These new clusters are created from the influence zones of the surface areas greater than the mean value (gray in Fig. 4.34d) or smaller than the mean value (white in Fig. 4.34d).

It is clear that this method of cluster detection leads to a radically different set of clusters than the method based on closing. However, the new approach, based on the size of influence zones can lead to new, interesting conclusions. We can state from Fig. 4.34d that smaller nodules tend to be closer to each other, whereas larger nodules tend to keep further away from each other. Obviously, this is only a very rough and qualitative analysis - a deeper insight can be obtained after the application of tools typical for stereology.[54,100]

Another type of colony-type arrangement can be often observed in high-speed steels (Fig. 4.35). In this material (and also in many others, usually precipitation hardened) we see two groups of carbide precipitates: those placed on the grain boundaries and those placed inside the grains. From the viewpoint of cutting properties or the tool's lifetime it can be interesting to answer the following questions:

- what is the proportion between carbides situated on the grain boundaries and these placed inside the grains?
- is there any relation between the size of carbides and the place they are observed?
- is there any clear relation between the size or positional distribution of the carbides and grain size?

In order to get any answer it is necessary to detect both grain boundaries and carbides. This is not easy, as the gray level of carbides and grains is the same (see Fig. 4.35a). Therefore, in this example we will devote more space to the description of the whole detection algorithm, starting from the gray image. The algorithm used here was as follows:

- the initial image (Fig. 4.35a) is not quite uniform. For example, we can note significantly darker regions in its lower part. Therefore, we start with a shade correction that leads to the corrected image shown in Fig. 4.35b
- the shade-corrected image can be a basis for binarization. There still exists some variation in the gray levels, therefore better results than after simple thresholding can be obtained after the use of a top-hat transformation. The structure was well etched, so we

get from the top-hat a well-defined binary image, with grains and, first of all, carbides detected as closed loops (Fig. 4.35c)

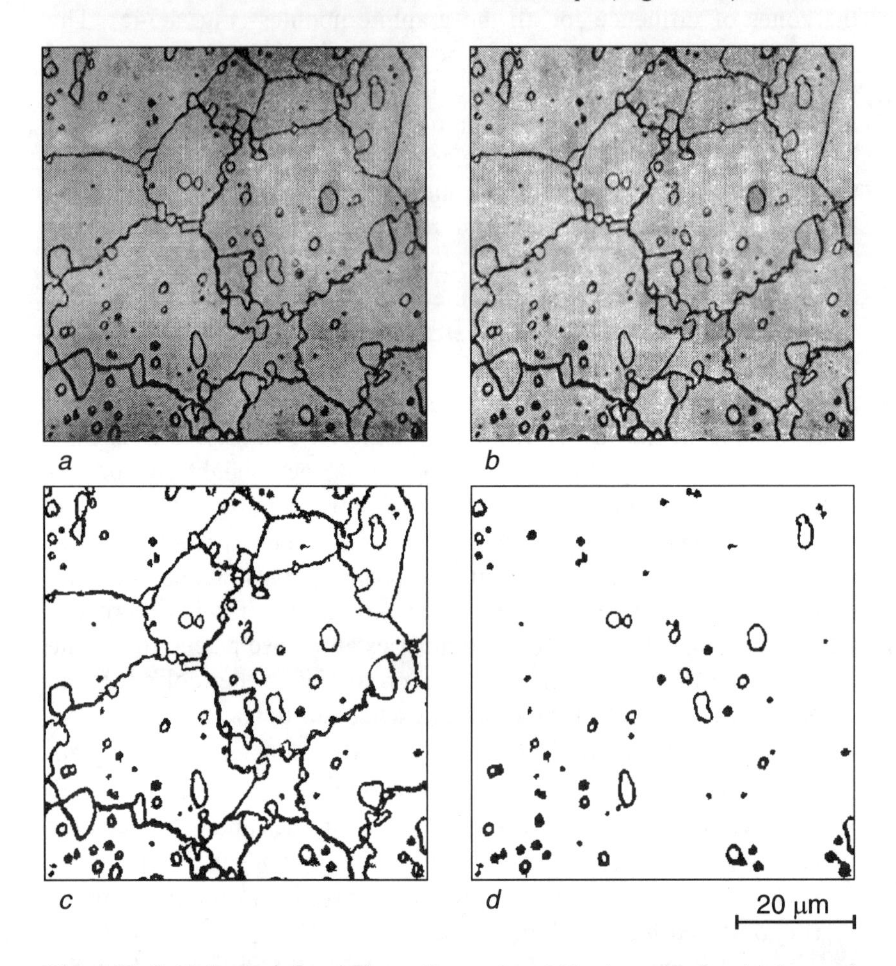

Fig. 4.35. Initial steps of grain boundary and carbide detection in a high-speed steel. Initial image (a), shade-corrected image (b), top-hat leading to a binary image (c) and internal carbides detected using the border kill procedure (removal of parts of the image cut by the image frame).

- the grain boundaries form a continuous network, so we can remove them, using the border kill procedure (removal of all the features cut by the image frame). As a result, we will get the carbides situated inside the grains (Fig. 4.35d)
- by simple subtraction of images from Figs. 4.35c and d, we get the image of grain boundaries and carbides intersecting these grains, as shown in Fig. 4.36a

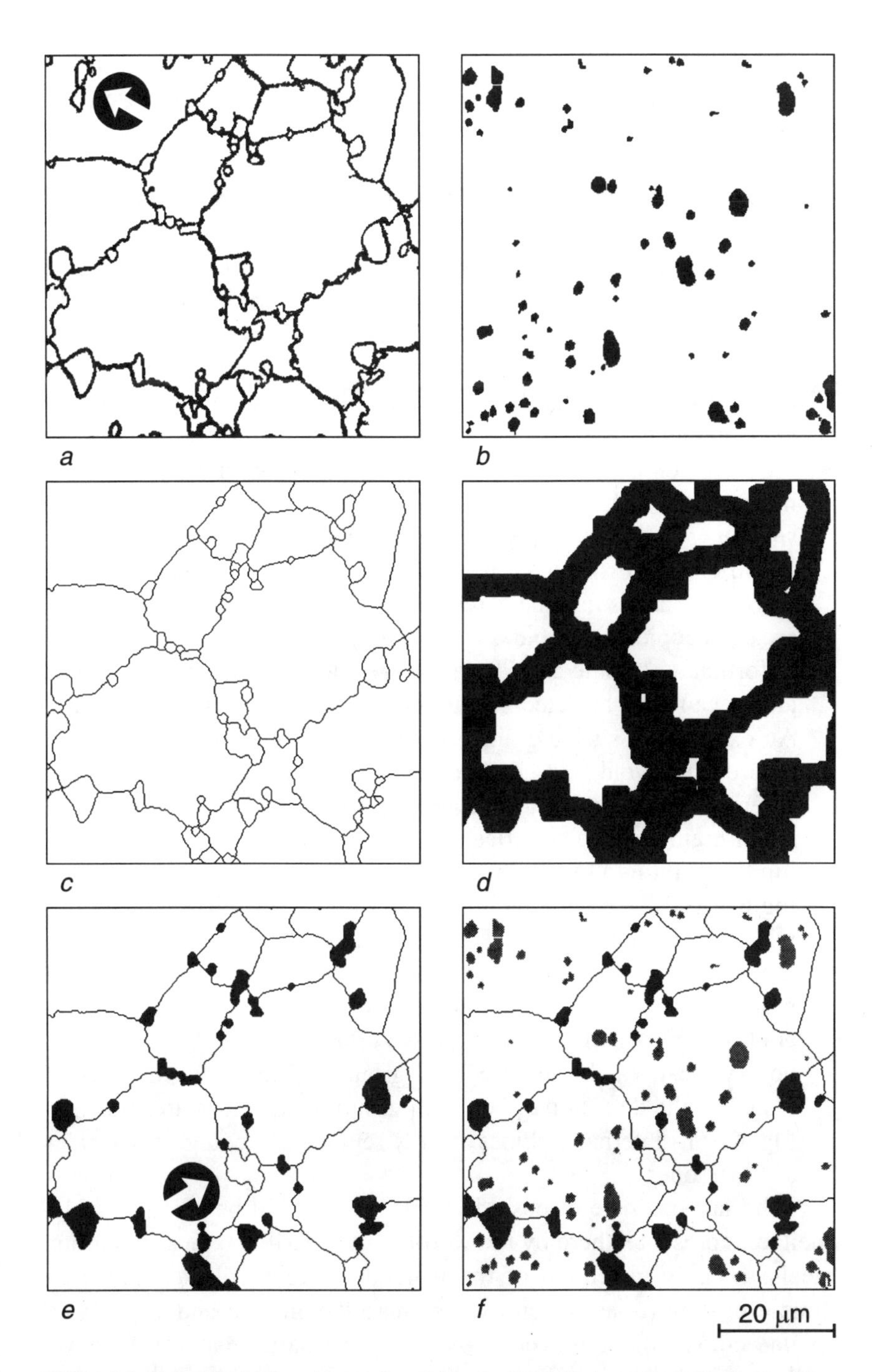

Fig. 4.36. Detection of grains and carbides - continued. Grain boundaries and carbides (a), internal carbides (b), single pixel-wide image of grains (c), eroded grains (d), carbides at grain boundaries (e) and the final detection (f).

- filling the closed loops in Fig. 4.35d will produce the final image of internal carbides (Fig. 4.36b). Note that in this image we have added the over-detected carbides indicated in Fig. 4.36a by an arrow. These carbides are detected by size-dependent filtering of features and subsequent addition to the previously detected internal carbides
- application of a SKIZ to the image in Fig. 4.36a allows us to create the single pixel-width image of grains and particles (Fig. 4.36d). Minimization of the width of the detected grains and carbides leads both to minimization of the detection errors during the final steps of analysis and to significant improvement in the accuracy of any measurements
- the next step is division of the preliminary detected image into two independent sets: grains and carbides. In general, carbides are smaller, so we will apply the sequence of erosion and further reconstruction. Erosion (Fig. 436d) will remove small particles (carbides) and subsequent reconstruction gives grains with black areas, occupied by carbides (Fig. 4.36e)
- unfortunately, some carbides are large and the procedure described above can lead to detection of some smaller grains like carbides. An example of such a grain is indicated in Fig. 4.36e by an arrow. In order to avoid this false detection, we should apply additional filtering (not shown in this sequence), based on the assumption that small grains can be significantly concave. This leads to transforming the grain into two or more sets during ultimate erosion. Next, a logical XOR step with the image from Fig. 4.36c will lead to grains with two or more holes. The next steps, skeletonization and removal of branches, produce loops glued together. Detection of multiple points, lying at the intersections of the loops, gives markers for reconstruction of the concave grains
- now we can easily analyze both groups of carbides. In order to visualize the detection effect, both groups of carbides are shown in Fig. 4.36f - internal carbides as gray regions and boundary carbides as black ones.

The last example discussed in this section will be colonies of lamellae, represented here by the well-known pearlite, typical of carbon steels. Usually, detection of the lamellar regions will succeed after simple closing (dilation can overestimate the amount and size of lamellae colonies). Moreover, if we apply a smaller magnification we will see the colonies as uniformly gray regions and we will have no need to use any special operation for detection of these colonies. By contrast, it will be difficult to detect the individual lamellae.

Lamellar regions (Fig. 4.37a, b) consist of a series of approximately parallel ribbons, being the traces of the intersection of the section plane and families of plates in the microstructure. Naturally, the ribbons are not endless and we always see some endpoints. These endpoints can define the grain boundaries of grains that were later transformed into lamellar regions. Sometimes the endpoints of the ribbons do not create a network of grain boundaries but only reflect some disorder or faults during formation of the lamellae. These regions can be called faulty regions and we will demonstrate below how they can be precisely detected:

- the computer program cannot detect precisely the endpoint of any ribbon as it has some width. Therefore, we have to change any ribbon into a single pixel-wide feature, reflecting the shape and size of the ribbon. The best transformation for this purpose is skeletonization. There are many algorithms for skeletonization. If your software offers any choice, you should choose the skeleton assuring homotopy. In most cases, it will be called L-skeleton. In rare cases we can also find a similar one, called M-skeleton.[21,84,114] The local irregularities in the ribbons produce some unnecessary, short branches. They can be easily removed using a few steps of a special HMT transformation called *pruning*. The pruned skeletons create a kind of symbolic representation of the lamellar structure (Fig. 4.37c)
- now, using another HMT transformation, we can detect the endpoints of the skeletons. These points describe the faulty regions, but if we want to visualize them, by dilation, for example, we will get large errors. Therefore, we will apply some additional, special treatment. For the pruned skeletons we will apply additional two steps of pruning. The resulting image will be added to the detected endpoints using a logical OR step. Consequently, we will get the image consisting of the pruned skeletons with the endpoints isolated by a single-pixel gap from the rest of the skeleton. If we now apply to this image a SKIZ transform we will get the small areas being places of faulty regions and larger areas being the influence zones of the shortened skeletons (Fig. 4.37d)
- reconstruction of this image, with the endpoints as the markers, will lead to detection of the faulty regions shown in Fig. 4.37e. We can check the final result of our analysis in Fig. 4.37f, where we have both the initial and detected images.

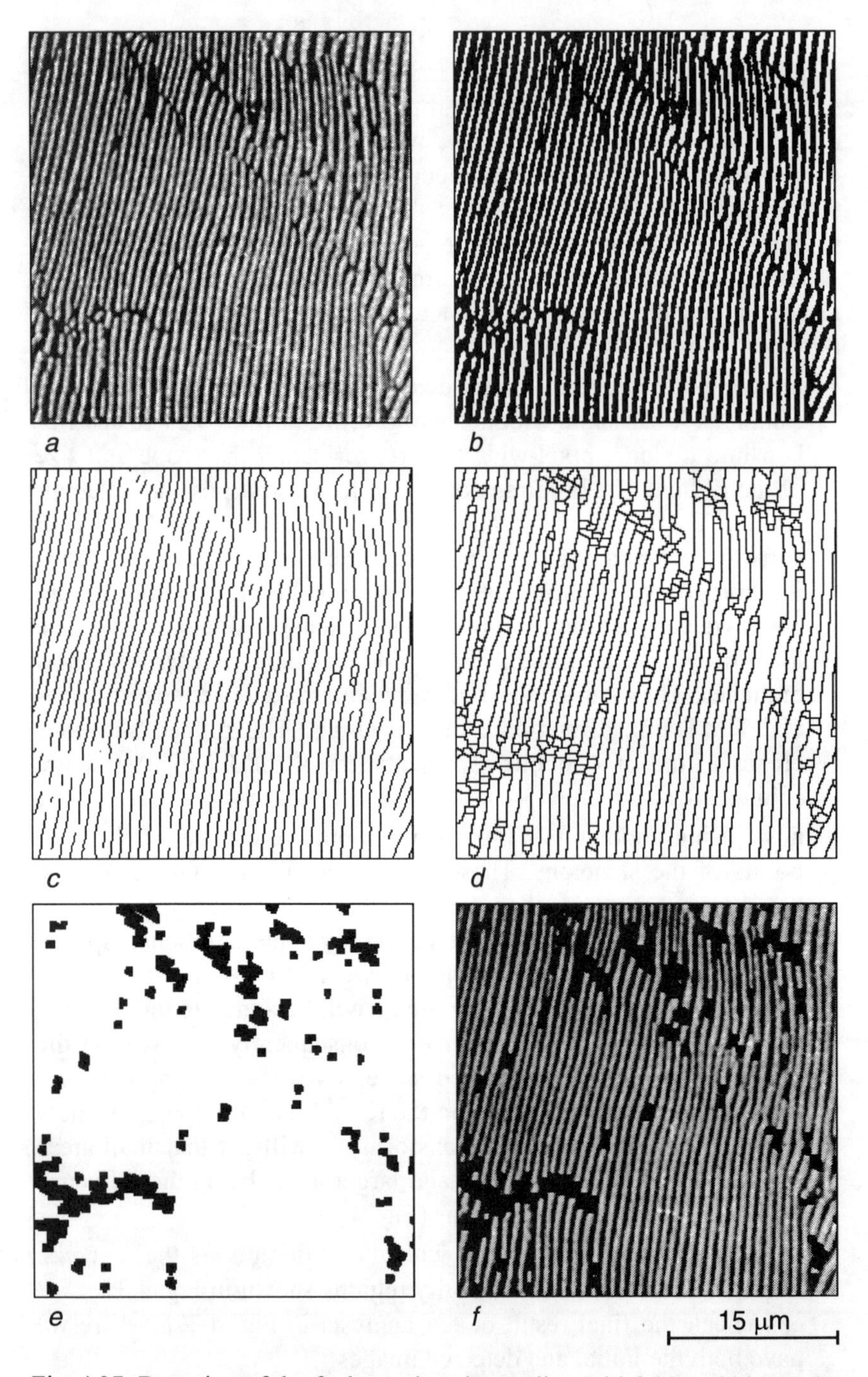

Fig. 4.37. Detection of the faulty regions in pearlite. Initial (a) and binary (b) images, pruned skeleton (c), SKIZ of specially prepared (see text for details) skeleton (d), faulty regions alone (e) and overlying the initial image (f).

4.5 Fibers

Fibers constitute the last group of basic features analyzed in this chapter. A fiber can be defined as a feature with one dimension significantly (usually at least one order of magnitude) larger than the remaining two dimensions. The fibrous character of the internal structure is found in various materials. The most typical examples, used by us every day, can be fabrics and cloths as well as many fiber-reinforced composites. Fibers are also present in materials of a natural origin, and wood is the most characteristic example.

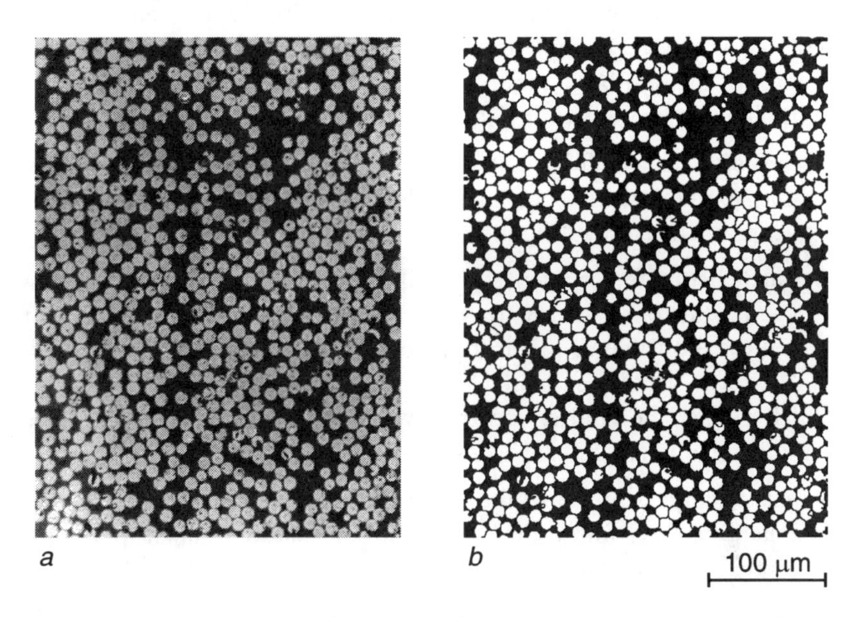

Fig. 4.38. A cross-section of a carbon fiber reinforced epoxy composite (a) and the same region after detection and separation of fibers (b).

The properties of fibrous materials are clearly related to the geometrical arrangement of the fibers. The best known property of these materials is that, in the case of parallel fibers, the mechanical properties are much greater in the parallel direction than in the direction perpendicular to the fiber orientation. However, in spite of a relatively simple description of the fiber geometry, the relationships between structure and properties of fibrous materials are very complex. Many subtle and difficult analysis factors, for example, the fiber-matrix interface, can affect the final properties. This does not alter the fact that the fiber properties and spatial arrangement are decisive for the final properties of fibrous materials.

The most convenient case for analysis is a cross-section of the material, oriented perpendicularly to the fiber direction as shown in Fig. 4.38. In the case of a good quality specimen, further processing is done without difficulty. For separation of fibers glued together, the methods described in Section 2.5 can be successfully adopted.

It is really difficult to properly detect and segment fibers parallel to the image plane, as in the linen fabric shown in Fig. 4.39. It is interesting that these fibers are quite easy to detect by a human, whereas it is enormously difficult for proper segmentation using computerized methods. In some cases it is possible only after very sophisticated treatment,[65] lying well outside the scope of this book. Here we will perform a relatively simple treatment, leading to detection of the basic properties of the fibers. First, let us try to detect the horizontal fibers in the linen fabric shown in Fig. 4.39:

- the initial image (Fig. 4.39a) is a little too noisy for segmentation due to the presence of many details produced by the inhomogeneous structure of the threads
- much better for segmentation will be the blurred image. We can use for this purpose any low-pass filter, even the simplest one, that returns the value of the arithmetic mean of the neighboring pixels. However, in this example a Fourier transformation and low-pass filtering of the Fourier spectrum were used, as they better preserve some characteristic properties of the image (Fig. 4.39b)
- any attempt to binarize the previous image leads to a very rough image, useless for further analysis. Fortunately, thanks to the directional illumination, we can find the axis of any fiber. For example, (see Fig. 4.39a, b), the central part of the horizontally oriented fibers is bright. This allows us to extract the axial part of the fibers with the help of gray-tone skeletonization (Fig. 4.39c). Gray-tone skeletonization differs a little from the binary case. In binary images the pixel is modified if its local neighborhood fits the structuring element. In the case of gray images, the transformation is performed if all the pixels at locations denoted by *1* are brighter than the central point and all the pixels at locations denoted by *0* are darker than the central point[114]
- binarization of the skeleton from Fig. 4.39c gives the binary image of the fiber axes. This image is a little distorted by unnecessary branches. In order to remove them we can perform pruning. Obviously, pruning will not destroy the axial lines creating closed loops (here of rectangular shape). The cleaned axes are shown in Fig. 4.39d

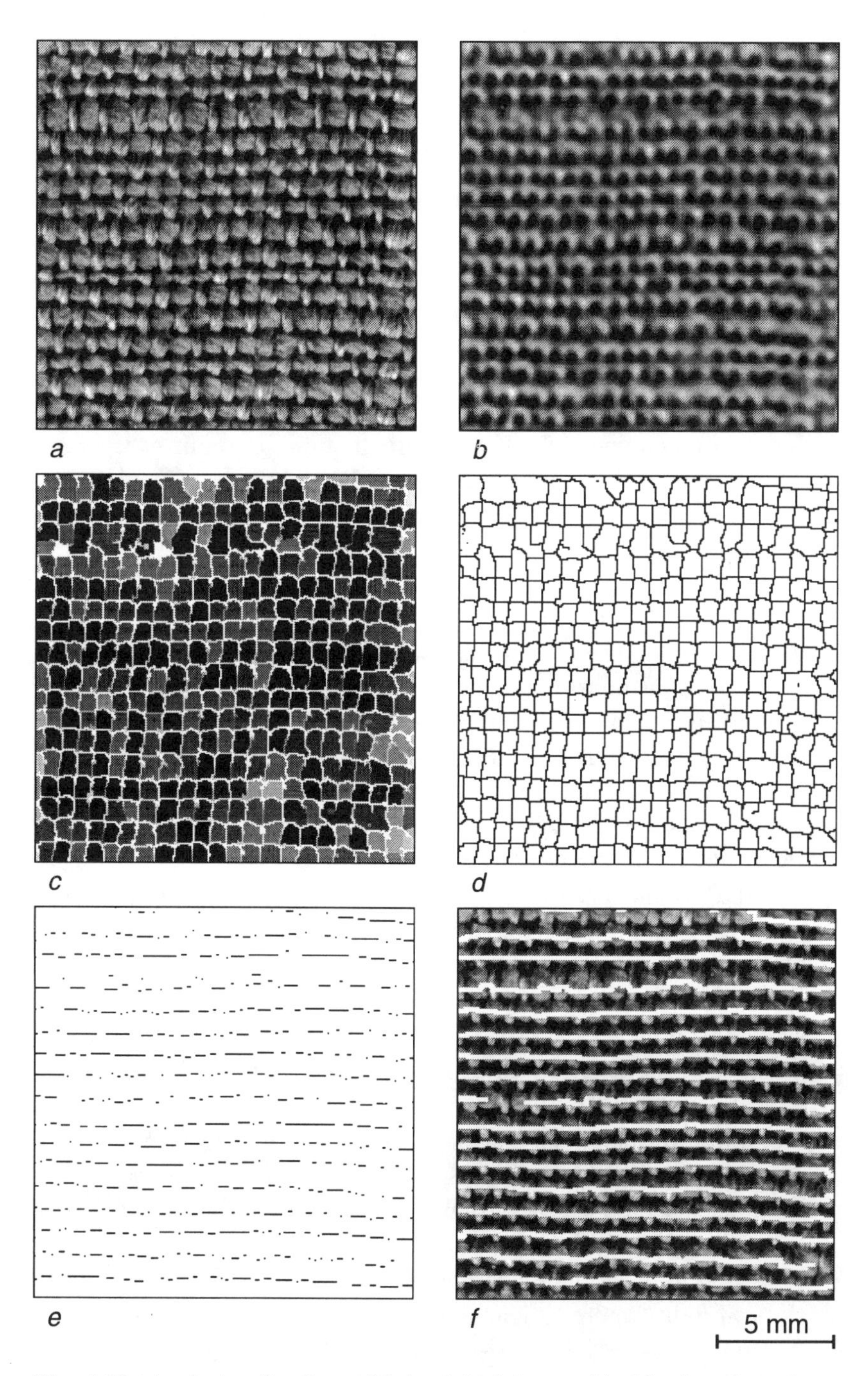

Fig. 4.39. Analysis of a linen fabric: initial image (a), blurring done by a Fourier transformation (b), skeltonization enhancing the fiber axes (c), binary form of the skeleton (d), detected horizontal elements (e) and final detection overlying the initial image (f).

- in the next step one should separate the horizontal and vertical fibers. Linear erosion allows us to detect the horizontal part (Fig. 4.39e) and subsequent linear dilation leads to restoration of the continuous axial lines of horizontal fibers
- the final image (Fig. 4.39f) illustrates the detected axes overlying the initial image. It allows visual control of the quality of detection, which seems to be satisfactory. The detected axes can be used, for example, for analysis of periodicity and accuracy during production of the fabric. A very similar analysis can be easily performed for vertical fibers.

The next example is devoted to the analysis of fiber distribution in a fiber-reinforced concrete. The initial situation is, in this case, completely different from that described in the previous example. The fibers are placed inside the opaque concrete matrix and are invisible. We can visualize them using X-rays that produce the initial image for our analysis (Fig. 4.40a). Now, we can process this image and one of the possible treatments is presented below:

- in the initial image (Fig. 4.40a) we see the traces of fibers visible as bright, thin lines. Obviously, it is impossible to reveal the full spatial arrangement of the fibers or even the number of fibers. However, it is possible to test the orientation of the visible components. Assuming a homogeneous distribution of fibers this will give sufficient information for analysis of the relationship between fibers and the properties of the concrete
- the background gray level in Fig. 4.40a is very inhomogeneous, mainly due to the concrete structure, which contains various gravel particles. Therefore, relatively good detection can be obtained using a white top-hat transformation (Fig. 4.40b)
- skeletonization and subsequent pruning leads to visualization of the axes of the readily visible, relatively long and thick fibers (Fig. 4.40c)
- the initial image can be converted into a Fourier spectrum. Appropriate low-pass filters will preserve only an oriented component of the image, which can be visualized after an inverse Fourier transformation (Fig. 4.40d)
- binarization of the previous image shows regions with high densities of bright features oriented in that direction (Fig. 4.40e)
- the logical intersection (AND) of images from Figs. 4.40c and d gives an oriented part of the detected fiber axes (Fig. 4.40f). Comparison with similar images, designed for other orientation angles, allows a full analysis of the fiber orientation, but obviously only in the image plane.

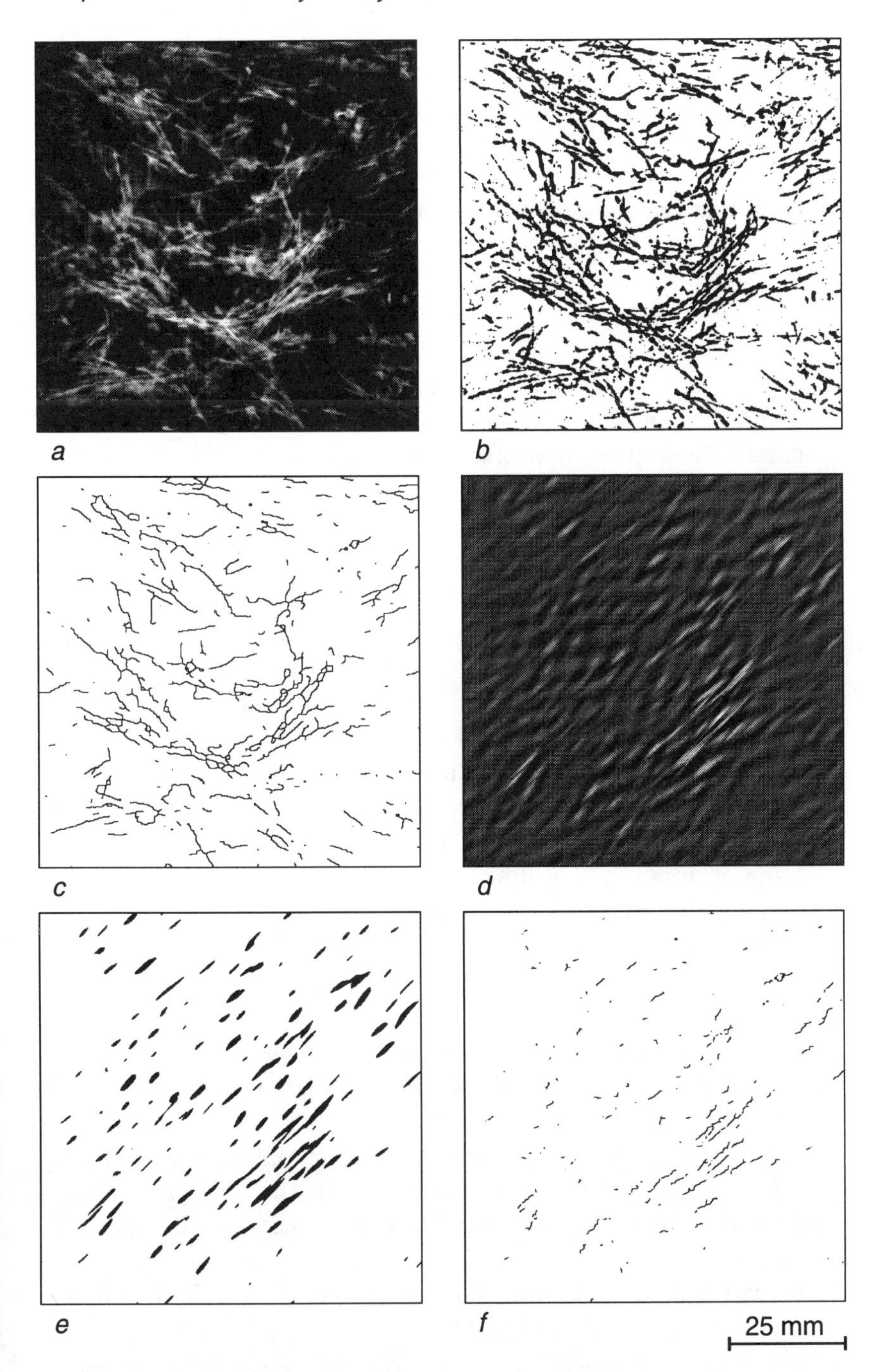

Fig. 4.40. Analysis of the fiber distribution in the fiber-reinforced concrete: initial X-ray image (a), fibers detected by top-hat (b), pruned skeletons (c), oriented component of the initial image in gray (d) and binary (e) form and final detection of oriented fibers (f).

Another type of structural component that can be treated as fibers are dislocation lines. Dislocations, being a local crystallographic defect, can be observed in thin foils. They require very high magnification, which is possible thanks to the achievements of transmission electron microscopy. The dislocation lines are relatively easy to see (Fig. 4.41a), but their detection is not easy. A suitable algorithm is described below:

- a heavy shading is clearly visible in the initial image (Fig. 4.41a). Its lower left-hand corner is much brighter than the remaining part of the image and this will disable any simple binarization. An attempt to apply a top-hat transform will significantly destroy the subtle dislocation network. Therefore, we should remove the shading prior to further analysis
- shade correction with subsequent removal of the noise by median filtering gives the expected homogeneous background (Fig. 4.41b). Unfortunately, the image still remains dark with poor contrast. An attempt to improve it by normalization gives no effect. A very small, isolated island of white pixels (indicated by the arrow) is responsible for this effect. The image contrast can be enhanced successfully by histogram equalization (Fig. 4.41c)
- although the dislocation lines are now clearly visible, they are still difficult for binarization. Depending on the threshold level we obtain either lines which are too thick with closed areas of smaller loops or discontinuous lines
- additionally, in spite of the equalized background, the gray levels of the dislocation lines exhibit large scatter. This can be improved by subtracting the equalized image from the same image, but after a dilation that removes the dislocation lines. As a result we get bright, thin dislocation lines. They can be subsequently enhanced by skeletonization (Fig. 4.41d), in a similar way to those in the previous example
- the next step will be binarization, which produces a good quality image of the dislocation lines (Fig. 4.41e). This requires some additional processing, as it is polluted by some very fine, accidentally detected features. The image can be successfully fine-tuned by partial pruning, leading to removal of very short lines, unnecessary branches or isolated points
- the above sequence of transformations produces the final image of the detected dislocation lines. This is shown in Fig. 4.41f, as an overlay of the equalized image. It is clearly visible in this image that the proposed procedure leads to a good quality extraction of the dislocation lines.

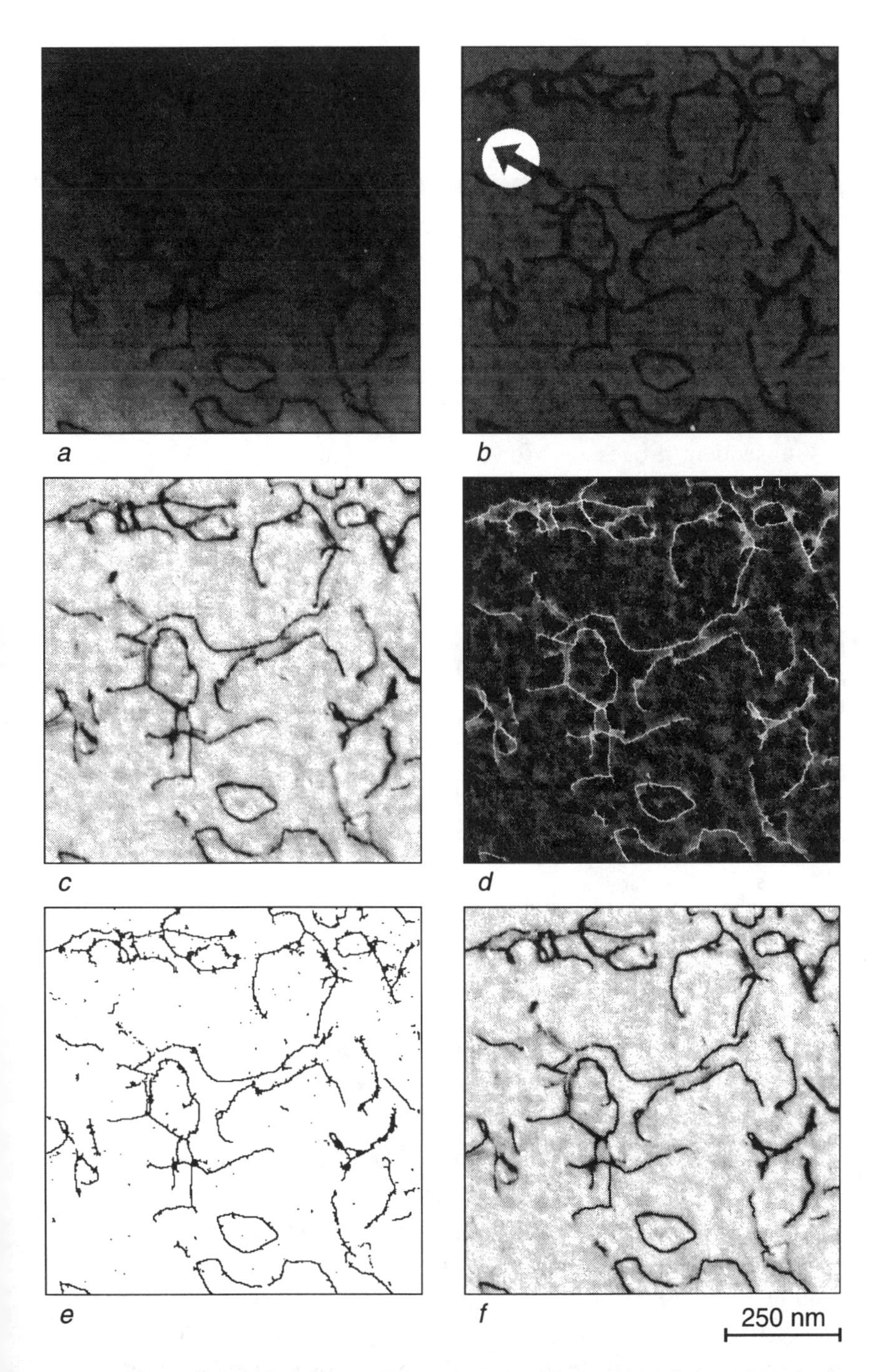

Fig. 4.41. A dislocation structure in Nimonic PE16: initial image (a), shade corrected (b) and equalized (c) image, enhanced dislocation lines (d), binary form of the skeletonized lines (e) and the final detection (f).

Now we will come back to the fiber-reinforced carbon-epoxy composite already mentioned at the beginning of this section. This time, however, we will analyze the fibers visible on fractures' surfaces. The first example (Fig. 4.42) shows the case when adhesion between the fibers and matrix is very good. This produces some traces of the matrix on the fiber surface (Fig. 4.42a). As a consequence, the fiber surface is very inhomogeneous, with many dark and bright spots, similar to noise. Let us analyze how we can process such an image:

- any attempt to threshold the initial image (Fig. 4.42a) leads to very poor results. We get a very noisy image, with fibers of various widths, partially glued together (Fig. 4.42b)
- subsequent processing, for example, opening (Fig. 4.42c), slightly improves the result of detection, but it is still not satisfactory. Especially in the upper part of the image we get discontinuous fibers that cannot be accepted as a result of image processing
- it was mentioned in Section 2.3 that median filtering can significantly reduce the noise level while simultaneously preserving relatively sharp edges. If we try this technique, we will get the result shown in Fig. 4.42d
- although the filtered image seems to be equally illuminated, it is only the first impression. Closer analysis proves that the middle of the image is significantly darker. Therefore, it is necessary to apply a shade correction (Fig. 4.42e)
- the shade-corrected image is suitable for binarization. We will have no difficulty in fixing the proper threshold level, as the background is homogeneous and almost purely black. A small opening will remove the traces of noise and we get quite a good result from detection (Fig. 4.42f). Its only drawback is some discontinuity in two fibers in the upper part of the image. This can be repaired by appropriate linear dilation.

The above analysis shows once more the difficulties in correct detection of fibers, already shown in the case of the linen fabric. Simultaneously it was shown that proper treatment allows correct and successful detection of the fibers. The process of solution discovery is similar in its principles to those widely presented in the case of grain boundaries. The main difference is in the characteristics of the objects to be detected: grain boundaries are thin lines dividing large areas, whereas fibers are relatively wide features occupying a large area of the image. Unfortunately, regions between fibers cannot be treated as boundary lines.

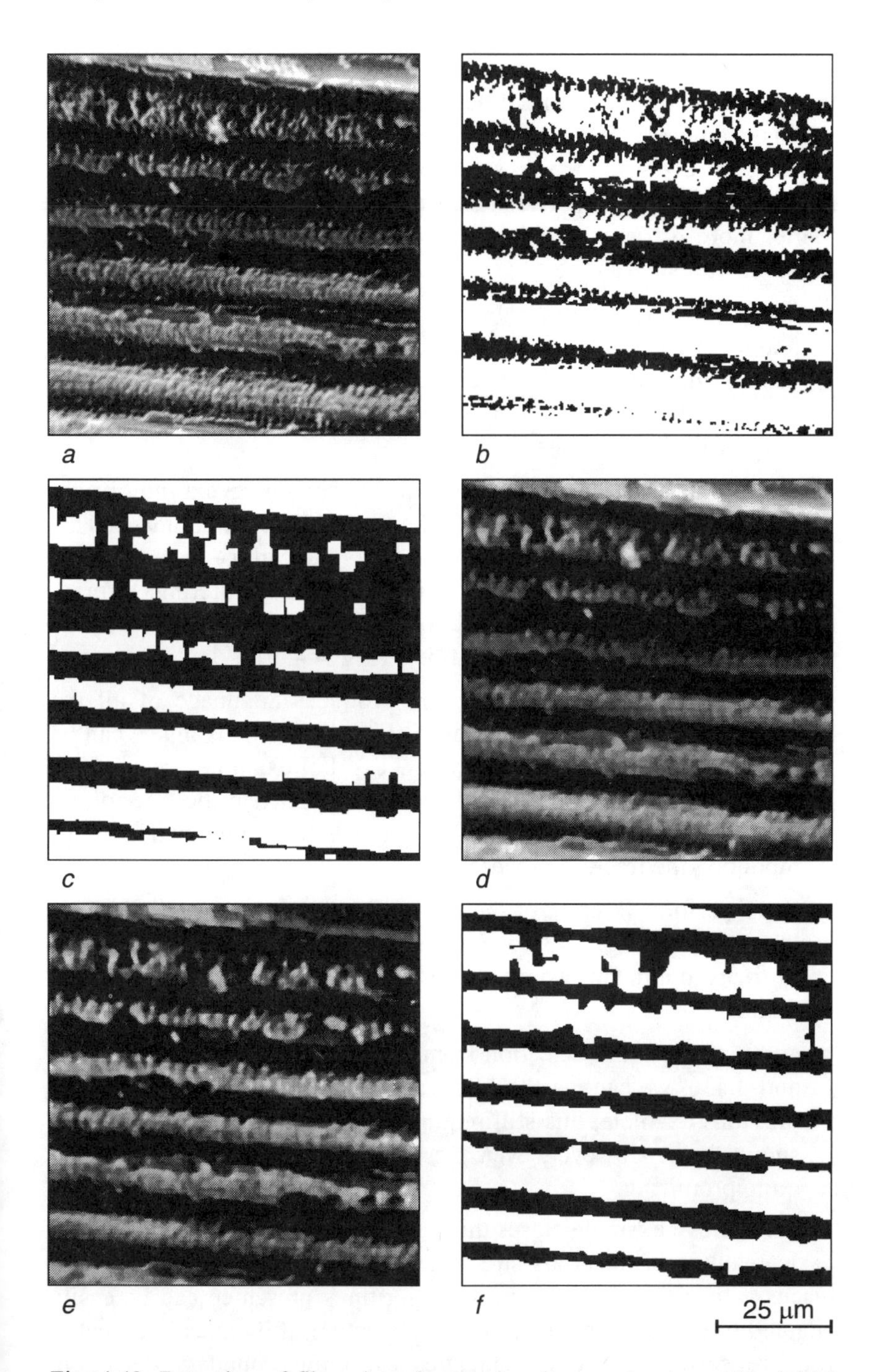

Fig. 4.42. Detection of fibers in a fractured carbon-epoxy composite: initial image (a) and attempts to detect the fibers by binarization (b) and subsequent opening (c), median filtering (d), shade correction (e) and final detection (f).

The last example presented in this section will also be devoted to the detection of fibers in a fractured carbon-epoxy composite. This case is even more difficult than the previous one due to a large scatter in the gray levels and the presence of bright bridges in the remaining matrix (see Fig. 4.43a). The presence of these bridges can effectively render more difficult any detection process. Let us analyze how we can deal with such a case:

- the initial image (Fig. 4.43a) is impossible for detection of any features by simple thresholding. On the other hand, however, it reveals some order: the fibers are oriented in one direction. This observation leads to an attempt to use a Fourier transformation as a tool for further analysis
- in the Fourier spectrum (Fig. 4.43b) we can observe a thin, bright line, oriented perpendicularly to the fibers in the initial image. This line is produced just by these fibers. So, we can try to remove the fibers from the image by proper high frequency filtering. In practice this can be done by removal of the part of the Fourier spectrum responsible for creation of the fibers (Fig. 4.43c)
- an inverse Fourier transformation produces an image of all the matrix bridges and fibers removed (Fig. 4.43d). This image can be used for analysis of the fracture process. For example, we can note that the matrix bridges exhibit some periodicity in the fiber direction. This can be further analyzed and is an important factor for modeling the fracture process
- if we perform filtering of the Fourier spectrum just the opposite to that shown in Fig. 4.43d, we will get an image of fibers slightly out of focus, but much easier to analyze than the initial image (see Fig. 4.43e)
- combining feature detection from the images shown in Figs. 4.43d and 4.43e, we can get the fibers as shown in Fig. 4.43f. As in the previous example, this still requires some additional treatment not shown here. However, even these preliminary results seem to be quite promising.

The above example closes this chapter, devoted to detection of the basic features in materials structures. Obviously, we were unable to discuss all of them. However, the algorithms presented can be easily adapted to other cases. For example, detection of dimples in a ductile fracture surfaces is almost identical to grain boundary detection. Therefore, deeper analysis shows that the material discussed in this chapter covers the majority of prevailing cases met in practice. Some more specialized aspects will be discussed in the next chapter.

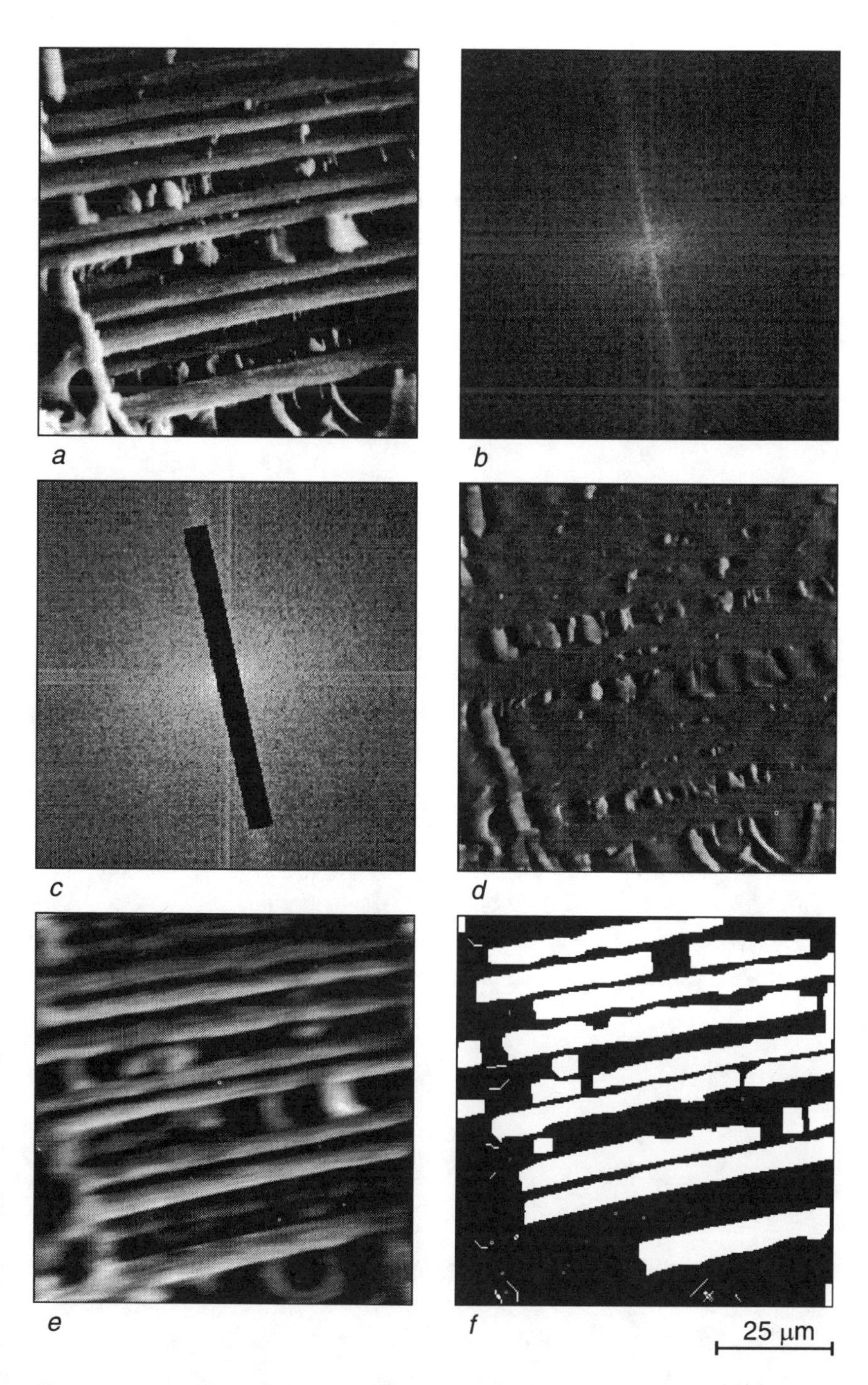

Fig. 4.43. Analysis of the fracture surface of a carbon-epoxy composite: initial image (a), its initial (b) and filtered (c) Fourier spectrum, inverse Fourier transformation (d), Fourier filtered fibers (e) and detected fibers (f).

Chapter five

Treatment of complex structures

Grains, particles and their aggregates, fibers or pores constitute the family of the most significant, basic and simple features, used to quantitatively describe the structure of materials. Chapter four was devoted to an analysis of these features. They can be treated as bricks for building, or rather analyzing and characterizing, the structure of the material. Clearly, the above list of structural components is highly subjective and can be easily increased. It covers, however, all the main components that seem to be responsible for the majority of mechanical, physical or chemical properties of materials.

Moreover, the features listed above are not rigidly defined and their characteristics can be easily extended. For example, procedures for grain boundary detection can be effectively used for the analysis of eutectic networks or the porosity of pre-pressed sinters. Similarly, procedures for fiber detection can be easily adapted to describe the network of dislocations or crack lines.

This chapter contains a little more advanced analysis algorithms. These are advanced in a sense, requiring either a more complex theoretical background or more powerful computational tools not available in all image analysis packages.

5.1 Textured and oriented structures

Texture, which is very difficult to define precisely,[80] is used here to describe image detail rather than preferred orientation. An illustrative example of this problem is shown in Fig. 2.27, which contains three types of simulated images. Every image requires entirely different segmentation methods:

- various features can be almost homogeneous and simple binarization is sufficient for their segmentation (Fig. 2.27a) or
- features can be separated using the more or less clearly visible and detectable boundary lines (Fig. 2.27b) or

- can be detected on the basis of different textures that are clearly visible but simultaneously difficult to define (Fig. 2.27c).

Detection based on different gray levels is very natural for almost all of us. For example, this text is detected by every reader thanks to the difference in gray levels between the paper and the print. This is obviously insufficient for recognition and understanding of the printed characters, but this has no meaning from the viewpoint of detection.

Separation based on different gray levels can be easily implemented using the threshold operation. In the case of textured structures binarization does not separate the objects (see Figs. 5.1a and b). This observation allows one to construct the descriptive definition of texture as a local property of the region in the image, connected with a variation in brightness or presence of any regular or irregular patterns. If we interpret gray levels as an elevation in a representation of the image as a surface, we will replace texture with roughness. The well-known difficulties with a precise definition of surface roughness explain the difficulties with defining the texture.

So, we can find in our images an endless number of different and difficult to define textures. Consequently, it is impossible to develop any universal tool suitable for texture-based segmentation. In practice, it is necessary to look for specialized transformations, sensitive to a given type of texture and to use them for segmentation. Fortunately, as will be shown below, the same texture can often be detected using various tools. Thanks to this we have a greater chance to find a satisfactory (however, not necessarily the best) solution.

It should be taken into account that the image texture discussed in this chapter is entirely different from crystallographic texture. The texture of an image can change dramatically with a change in image resolution. For example, a textured, macroscopically homogeneous material can be revealed as a collection of particles embedded in a matrix, if only observed at sufficiently high magnification.

The above remarks show how complex and difficult analysis of textured images is - we have to detect features that are only approximately defined and tend to change with magnification or other observation parameters. Consequently, we rarely have a chance to fully and objectively verify the quality of our detection. Therefore, in this chapter detection of textured components is performed on simulated images. The test square image (Fig. 5.1a) consists of three vertical bands of thickness equal to 0.25, 0.5 and 0.25 of the square side, respectively. The outer, thinner bands have the same texture. The inner, thicker band exhibits another texture. The textures are difficult to distinguish by a human observer.

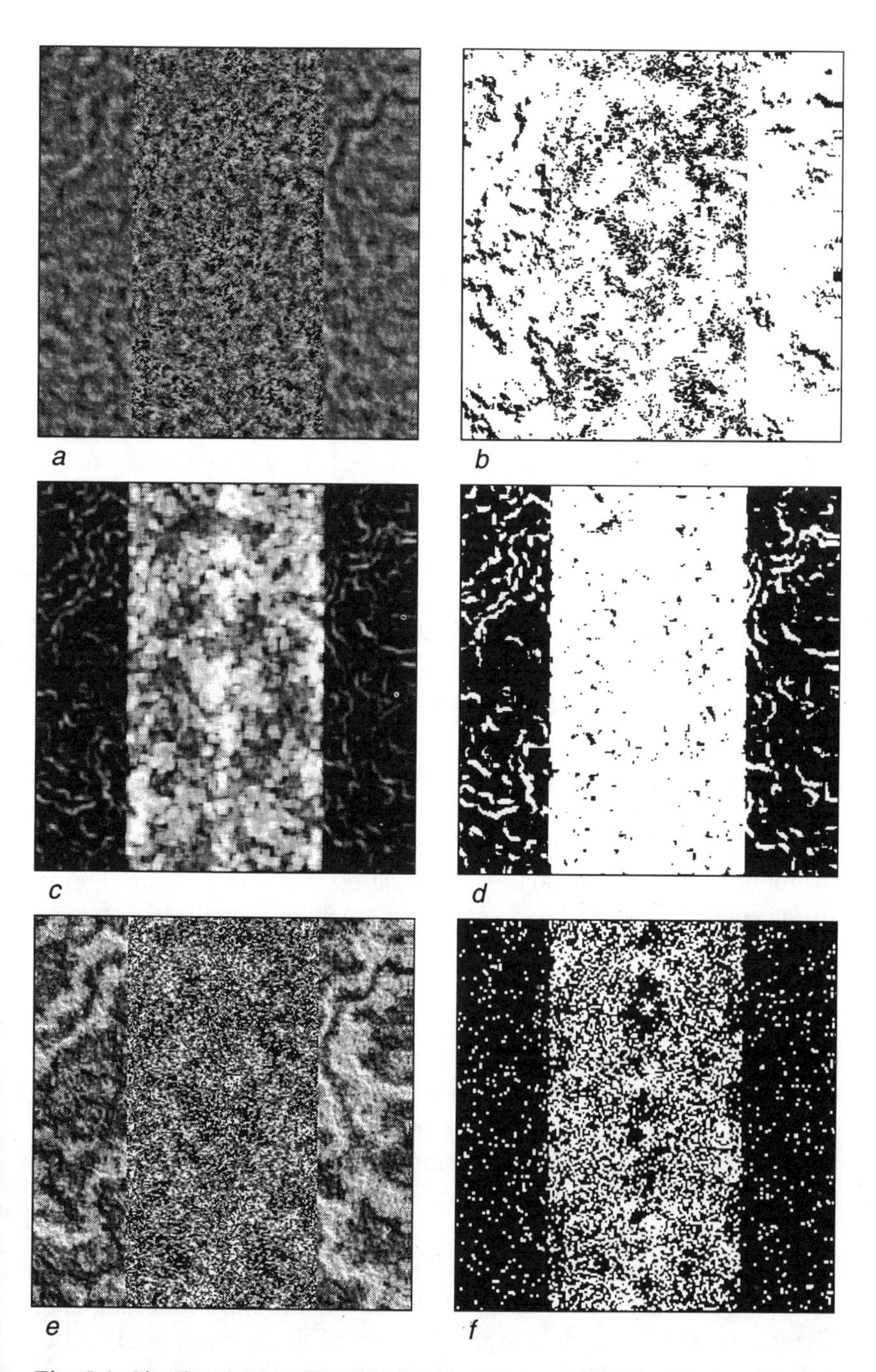

Fig. 5.1. Simulated structure with different textures (a). Binarization does not detect the features (b). Good results are obtained from the local variance (c) with subsequent thresholding (d). Acceptable results are given also by sharpening (e) with subsequent thresholding (f).

A method for detecting the features in this simulated structure, as well as the difficulties in this process, is illustrated in Fig. 5.1. It can be noted from the initial image (Fig. 5.1a) that both textures give the same average value and overall scatter of gray levels. Consequently, they cannot be separated by thresholding. This leads only to a set of black spots and isolated points that cannot be used for assessment of the boundaries between regions with different textures (Fig. 5.1b).

However, the textures differ in the local arrangement of pixels. The outer texture seems to be smoother, with lower local contrast, than the internal texture. Therefore we can try to compute the *local variance* of gray levels in neighborhood of any pixel. Indeed, the values of variance significantly differ for both textures (Fig. 5.1c). This last image can be easily converted into a binary form, using a simple thresholding procedure (Fig. 5.1d). The binary image requires additional final processing in order to remove small, black or white, features. This can be done using various methods. The simplest solution seems to be the application of closing and subsequent opening procedures, leading to removal of the small black and white features, respectively.

It is expected that the qualitatively noted higher local contrast in the inner texture should be emphasized during the sharpening process. We can apply a sharpening filter, as described in Chapter two and shown in Fig. 2.7b. The simulated, textured structure after sharpening filtering is shown in Fig. 5.1e. The sharpened image still looks to be an impossible segmentation by binarization, but thresholding returns an acceptable result (Fig. 5.1f). After relatively simple fine-tuning operations (for example, median filtering) we can produce from this binary image a correct final result.

Now we will try to use another technique, based on *correlation.*[80] From the initial image (Fig. 5.2a) we cut a small piece (marked in the upper left-hand corner by a square frame) which will act as a template for further analysis. Correlation is performed by sliding the template kernel over the input image and computing the image response at each position. The most popular methods of calculation of this response are based on subtraction and multiplication, respectively.[31]

Subtraction-based correlation (Fig. 5.2) depends on subtracting the corresponding pixel values of the analyzed image and template. The image response is calculated as a sum of abstract values of all the differences. So, in the case of perfect matching the response is equal to zero, whereas for any mismatching we get values greater than zero; the greater the bias of matching, the greater the response value. This last property is clearly visible in Fig. 5.2c. The regions of good matching of the template are darker than the middle of the image, occupied by another type of texture. Binarization of the correlation image gives approximate division between two types of texture in the initial image (Fig. 5.2a).

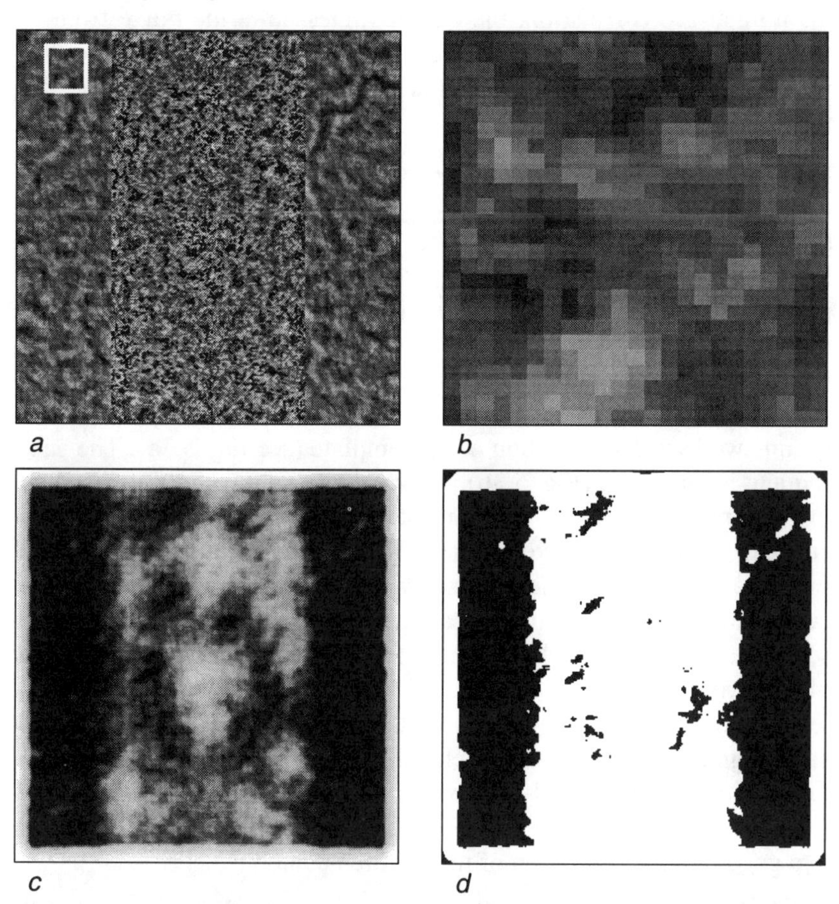

Fig. 5.2. Detection of textured components: initial image with the highlighted template (a), enlarged template for correlation (b), correlation image obtained using the subtraction technique (c) and separated textures in a thresholded correlation image (d).

A considerable disadvantage of the correlation technique is the fact that it gives erroneous values at the image edges. In the case of Fig. 5.2 this effect is visible in the form of a bright frame in images 5.2c and 5.2d. In order to avoid or minimize this error, the image should be padded with zeros. Slightly better mathematical precision can be achieved by padding the image with the average values of the original image brightness.[80]

Correlation based on multiplication is fully analogous to the previously described variant based on subtraction. Additionally, the result can be normalized, enabling more quantitative, automatic interpretation of the results of the transformation.[80] In the case of application of the normalized correlation, the output values lie in the range from -1 to 1, with 1 and -1 indicating perfect (anti)-correlation. The value of 0 indicates no correlation between the image data and the template pattern.

An example of this type of detection is shown in Fig. 5.3. Obviously, as we cannot see the pixels with negative gray levels, the results of transformation have to be visualized using the appropriate LUT transformation. Therefore, pixels with the value of 0 are displayed as 50% gray, whereas the pixel with the value of -1 or 1 is displayed as black or white, respectively. Perfect matching (the value of 1) gives the brightest point in the correlation image (indicated by an arrow in Fig. 5.3a). Note that the location of this white point corresponds well with the location of the template (see Fig. 5.2a). The subsequent binarization (Fig. 5.3b) is done with two threshold values: the lower threshold cuts off the darkest pixels and the upper threshold cuts off the brightest pixels. The detected area is built of the pixels with the local environment significantly different (according to the correlation technique) from the template. The resulting binary image seems far from correct detection, but can be easily improved by simple filtering. A small opening will remove all the narrow bridges, leading to a large region in the middle of the image, surrounded with some significantly smaller particles (Fig. 5.3c). The central part of the image can be significantly improved by hole filling (Fig. 5.3d). The final transformations, opening (Fig. 5.3e) and closing (Fig. 5.3e), lead to a good quality separation of the textured components.

The above examples show that correlation techniques give only approximate results. Error of detection (in sense of location of the boundary line) is roughly proportional to the template size. On the other hand, however, correlation seems to be a very efficient tool if we have no indications how to differentiate the textures. Another application of the correlation techniques is a search of small, precisely defined components of the image.

Textured images can be also segmented using some techniques of automatic thresholding. We will use for this purpose a binarization method based on the local histogram.[4,50] The principles of this technique are briefly described below, whereas an example of its application is shown in Fig. 5.4.

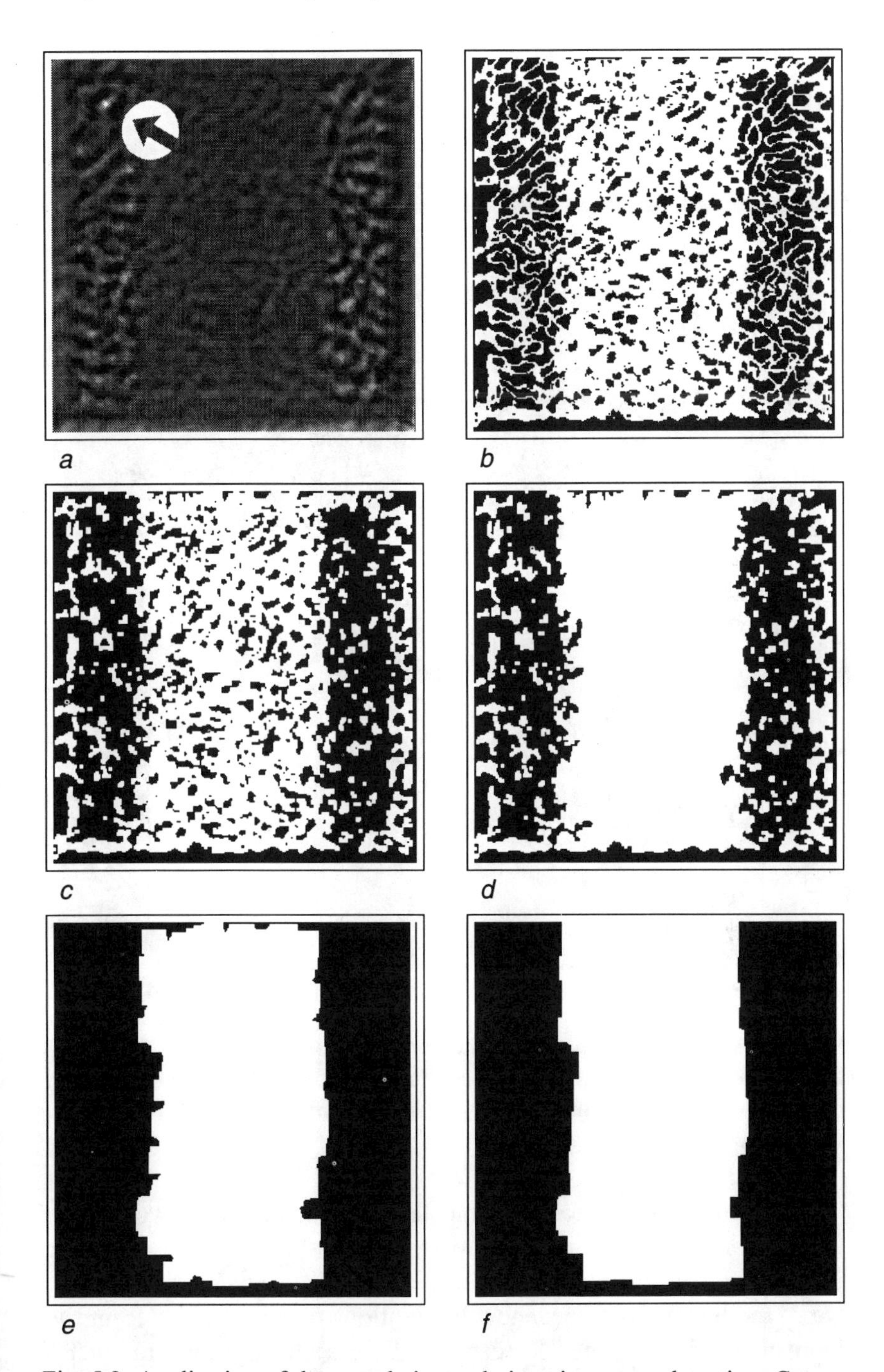

Fig. 5.3. Application of the correlation technique in texture detection. Correlation image with the brightest point localized at the same place, as the template (indicated by an arrow) (a), binarized image (b), opening (c), hole filling (d), final opening (e) and closing (f).

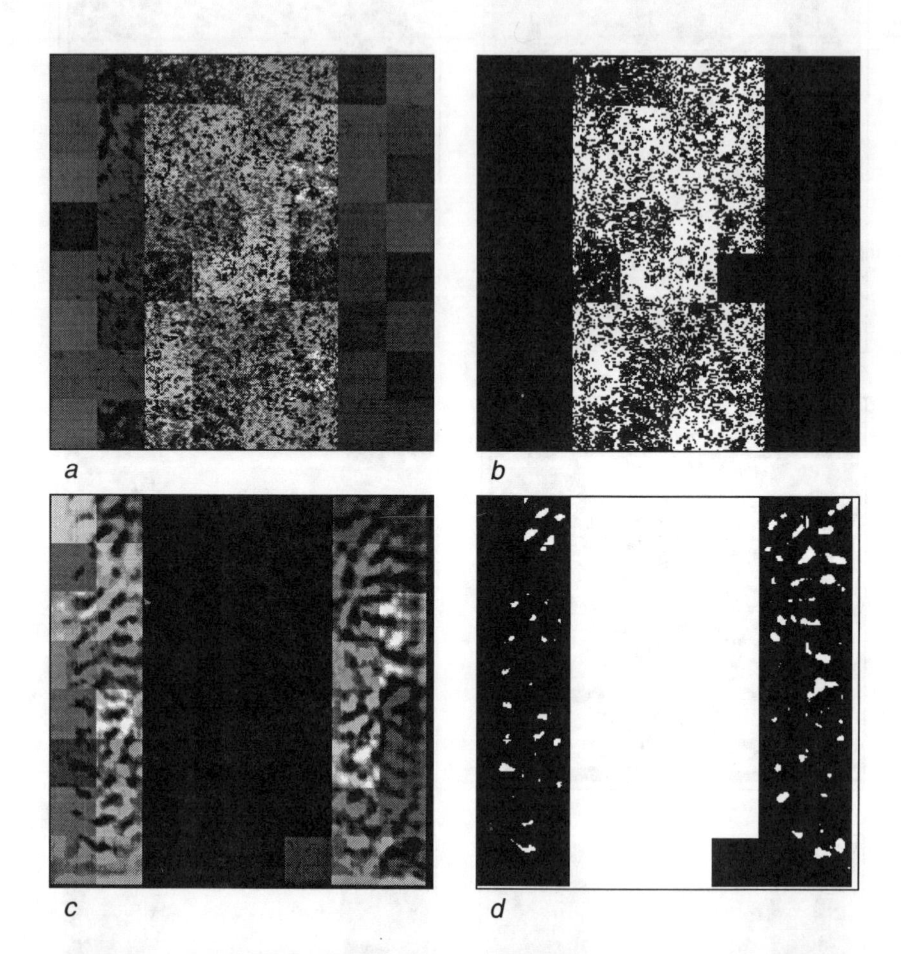

Fig. 5.4. Application of the local histogram threshold in texture analysis. Preliminary detection (a) and binary image (b) applied to the texture image from Fig. 5.1a. Same preliminary detection (c) and binary image (d) applied to the correlation image from Fig. 5.3a.

A local histogram-based segmentation is performed in two phases. In the first phase, the image is partitioned into a set of rectangular local sub-images called sectors (in Fig. 5.4 we use 8x8 = 64 sectors). Next, the histogram (frequency distribution) of pixel values from the input image is computed for each of these sectors. Within the histogram of each sector significant clusters are identified by means of the peak-valley analysis.[4,50]

The second phase of the process involves the peak addition in which ambiguous peaks in the histogram can be verified according to their presence or absence in the adjacent sectors. The intensity value of each of the selected peaks is then used as the output label for all pixels in the input sector that map to the corresponding peak. The domain of this mapping includes all intensity values lying between the valleys on either side of the peak.

An attempt to apply this sophisticated algorithm to texture detection and analysis, shown in Fig. 5.4, proved its usefulness. Similarly, as in the case of correlation, the results are usually only rough and approximate.

The last technique applied in this chapter for texture analysis will be the Fourier transformation. It is one of the most flexible and useful methods in image analysis (examples of its application are also presented in previous chapters). One should, however, use the Fourier analysis with great care as it can easily produce completely wrong results. Moreover, it requires a lot of computing power to get results in a reasonable time.

The same test textured image (Fig. 5.5a) as in the previously described methods is used for analysis. The Fourier spectrum (Fig. 5.5b) is filtered using a high-pass filter. The filtered spectrum is shown in Fig. 5.5c and it is clearly visible from this image that all the low-frequency information was thrown away. The inverse Fourier transformation produces an image containing only very subtle relief produced by the highest frequencies of the initial image. This relief is so delicate that it is hardly visible. In order to make it visible to the human eye, the intermediate image produced by the inverse Fourier transformation was enhanced using the brightness/contrast control. The enhanced image is shown in Fig. 5.5d. The subsequent top-hat transformation (Fig. 5.5e) and closing of the binary image lead to correct detection of differently textured areas (Fig. 5.5f). Closing of the holes visible in Fig. 5.5f will produce an almost perfect detection.

To summarize the texture analysis presented in this chapter, we can state the following practical remarks:

- texture analysis is difficult, and can easily lead to false results. Perfect results are rarely achieved, much more likely is approximate detection (i.e., containing some errors)
- correct texture detection requires high sensitivity to small changes in local collections of pixels. The consequence of this fact is that texture analysis is not very suitable for fully automatic analysis
- probably the main problem with elaboration of suitable algorithms lies in the difficulty of objective verification of the quality of the detection fulfilled. This was the main reason for the use of a simulated structure in the analysis presented in this section.

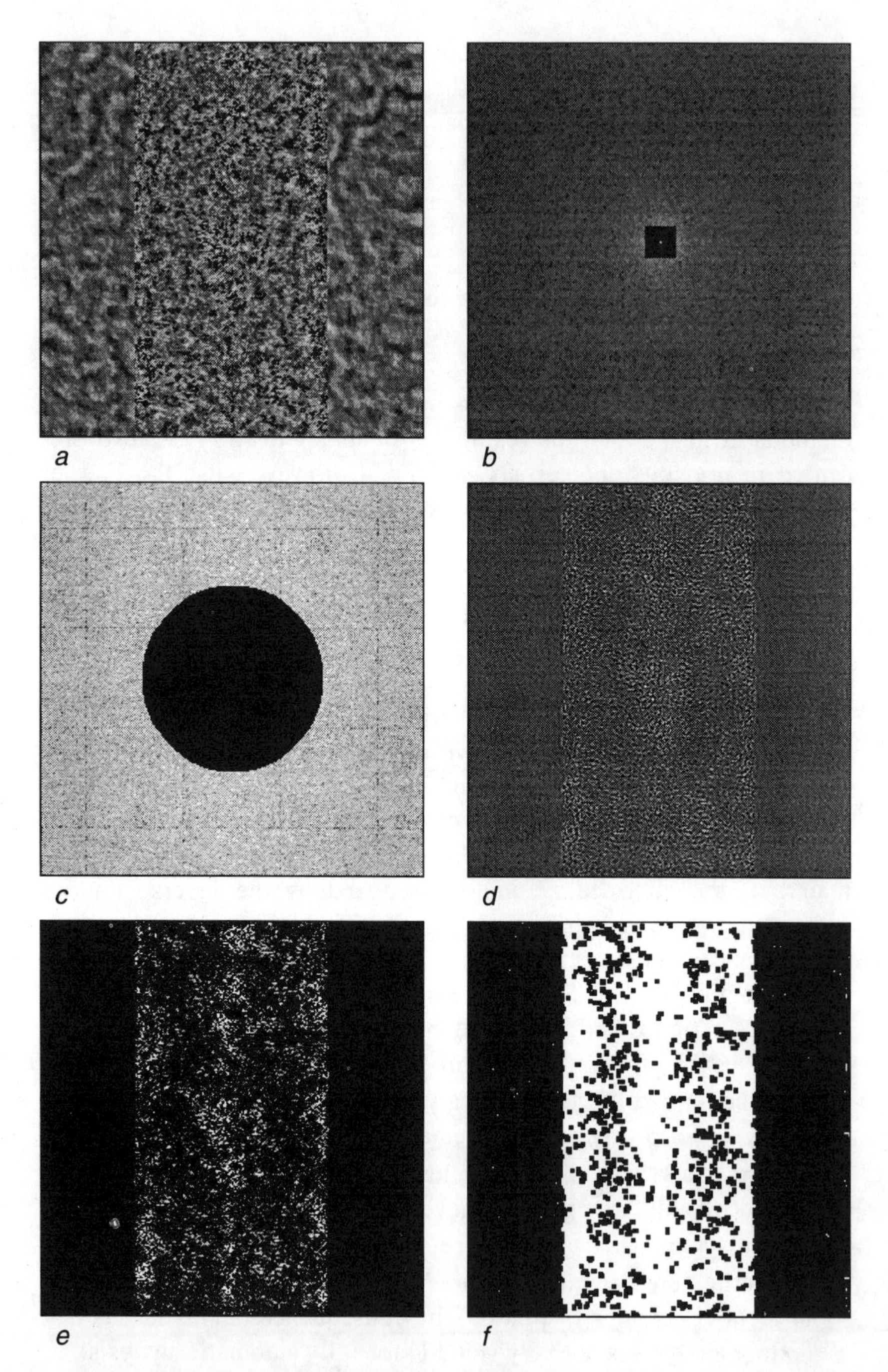

Fig. 5.5. Application of the Fourier transformation in texture detection: initial image (a), its Fourier spectrum (b), the same spectrum after high-pass filtering (c) and subsequent inverse Fourier transformation (d). Top-hat-based detection (e) followed by binarization and closing gives the final result (f).

5.2 *Very fine structures*

Some materials have extremely fine internal structures. Nanomaterials, metallic glass or precipitation hardened materials can be listed as illustrative examples. Investigation of their microstructure often requires the use of transmission electron microscopes (TEM) which offer very high magnifications. Usually specimens for TEM are prepared in the form of thin foils and this makes this technique difficult and laborious. Natural irregularity of the foil thickness significantly affects both the image quality and a number of possibly overlapping features. Finally, the small size of structural constituents means that even at the highest magnifications they are represented in the image only by a few pixels.

All the factors described above make the analysis of very fine structures especially difficult. The basic methods of image analysis, based on such transformations as thresholding, edge detection or simple morphological transformations are usually useless. Much better results can be obtained if we process the images as textured structures. However, as was shown in the previous section, texture analysis gives only approximate results in the majority of cases.

The absolute majority of materials reveals clear and evident crystallographic structure. Traces of this can be found locally even in materials recognized as amorphous, like plastics or metallic glass. The main properties or crystallographic structures are: order and periodicity. Additionally, the smallest dimensions of the crystal network are only two orders of magnitude greater than the electron wavelength in TEM. As a consequence, HREM (high-resolution electron microscopy) images are very sensitive to any interference phenomena which can locally affect the image contrast in an unexpected way, making further analysis even more difficult. On the other hand, however, this property opens new perspectives for image enhancement and analysis techniques.

The most promising tool for treatment of the images described above seems to be Fourier analysis. There are at least three major areas of application of Fourier analysis:

- enhancement of the image, based on filtering of the lowest frequencies in the Fourier spectrum. This can lead to restoration of previously invisible details and overall improvement of the image contrast (see Fig. 5.6)
- detection of various phases or structural constituents within the image, as shown in the previous section, devoted to texture analysis (see Fig. 5.5)

- identification of structural components, based on the analysis of diffraction patterns[77] (Fig. 5.7).

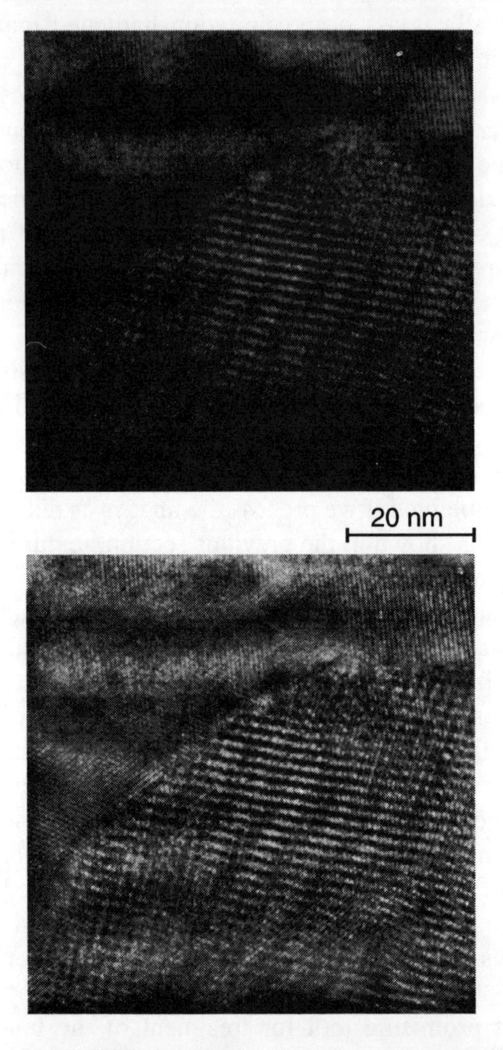

Fig. 5.6. Application of Fourier filtering in enhancing an HREM image. The initial image of $Y_4Al_2O_9$ oxide dispersoids in a ferrite matrix (top) after Fourier transformation, filtering out the low-frequency component and inverse Fourier transformations observation of the fine structure in the matrix (bottom).

Diffraction patterns are widely used for identification purposes in TEM.[27] Unfortunately, even this extremely efficient tool cannot be applied for analysis of very small areas of a few nanometers size. In this case we can effectively apply the Fourier transformation.

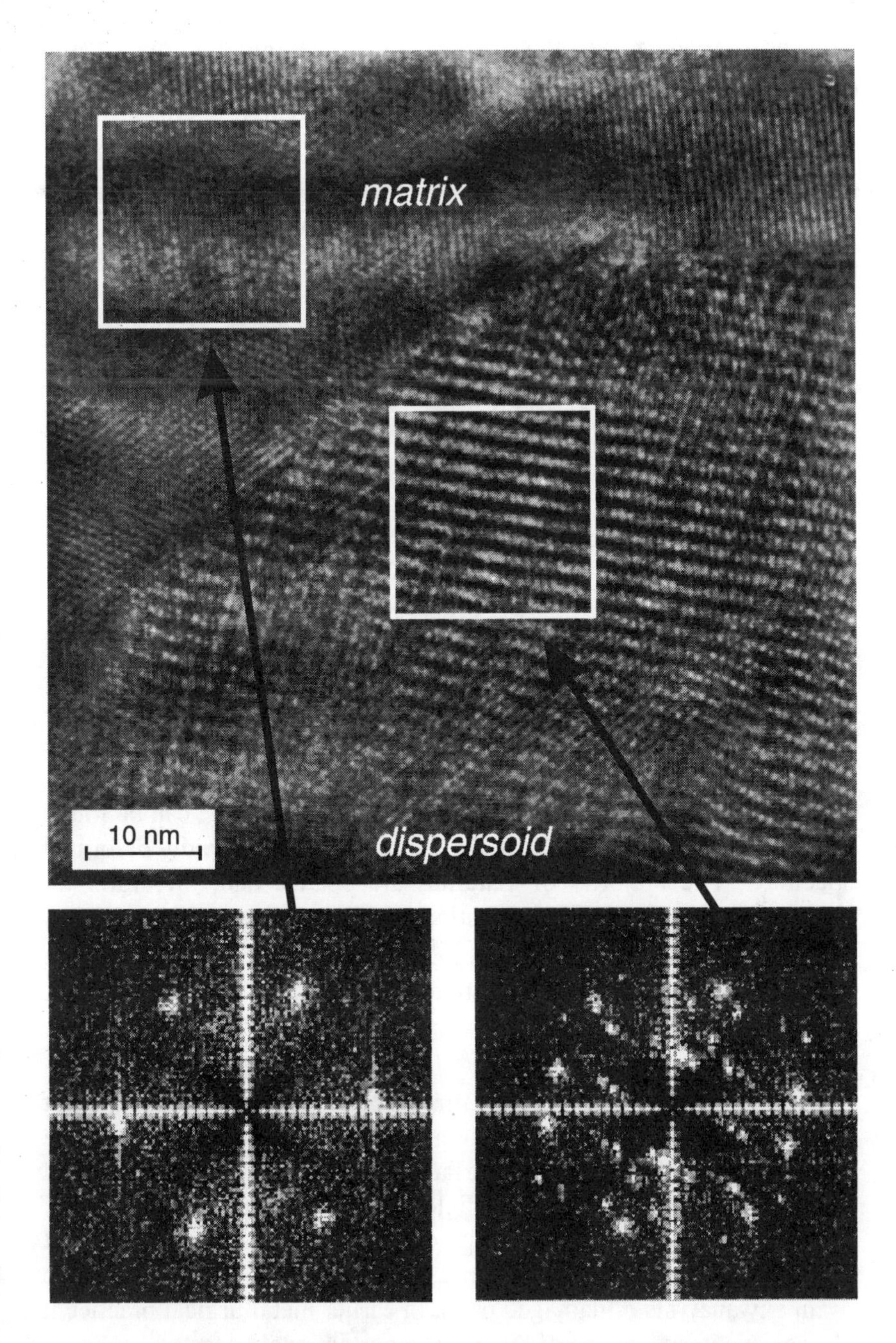

Fig. 5.7. Application of a Fourier transformation in the identification of the phases in $Y_4Al_2O_9$ oxide dispersoids in a ferrite matrix. HREM image (top) and the diffraction patterns obtained from extremely small areas of size 15x15 nm (bottom). The quality of the diffraction patterns is sufficient for identification purposes.

The HREM image shows clearly the crystallographic planes (see Figs. 5.6 and 5.7). Therefore, using the Fourier transformation we can effectively simulate creation of the diffraction patterns typically used for identification in TEM. Additionally, we can apply any other image processing tools in order to filter and enhance the image. Final computations, leading to recognition of the composition of a given region, require specialized software[118] which is beyond the scope of our analysis.

5.3 Fracture surfaces

The fracture process is very difficult for any experimental research. Obviously, we usually cannot predict and, consequently, observe unexpected, catastrophic fracture events. But to make things worse, we also cannot do it on a laboratory scale, either. The fracture process is extremely sensitive to changes in the local stress and strain state. Under laboratory conditions we can only observe the crack behavior at the specimen surface, corresponding with the plane stress and three-dimensional strain state and it is insufficient. It was proven both theoretically and experimentally that the decisive factor for the fracture behavior of the material is the plane strain state (it means simultaneously the three-dimensional stress state). This stress state can be found only in the vicinity of the crack tip inside a thick and rapidly deformed specimen. The majority of materials are opaque and, therefore, we have no chance for direct observation of this process.

As a consequence of the above described properties of the fracture process, the only available solution is analysis of fracture surfaces. There are numerous reasons for which analysis of fracture surfaces is so important:

- fracture process is highly localized - this reflects well the concept of the weakest chain link, responsible for the strength of the whole chain. Therefore, fracture surface is *the only place* we can find structural phenomena responsible for a given fracture behavior of the material. Thus, fractographic examination is always the first (and often decisive for the final result of the whole analysis) step in any analysis of damaged parts. In earlier metallurgical practice it was even used as one of the most important quality tests
- if we look for the reason of any failure caused by fracture, we try to check experimentally the conditions of crack formation and growth. If two separated fracture surfaces are geometrically very similar, we can assume that the local conditions of fracture forma-

tion were also very similar. In order to compare these fracture surfaces in an objective way, we need to define some measures - this is one of the roots of so-called quantitative fractography.[30,93,101,107] It should be stressed that quantitative fractography has its own weaknesses but is the best known tool for analysis of the crack formation process

- many structures are intentionally overloaded. The most typical example can be the aircraft wing structure where we tend to keep the construction as light as possible. In such constructions the fracture process is predicted as a normal consequence of the exploitation process. Therefore, during design of the whole structure various fracture resistance tests are performed in order to assure safe operation of the final product. Careful, quantitative analysis of fracture surfaces is in this case very helpful for evaluating the crack growth rates, crack initiation conditions, etc. (Fig. 5.8).
- during design of an entirely new material or optimization of an already existing material, fracture properties are among the most significant factors. This can be easily understood if we note that usually, when increasing the strength of the material, we decrease simultaneously its plastic properties as well as its fracture toughness. Precise analysis of the fracture surfaces gives a lot of information concerning the material's response to the failure load. The importance of such analysis was established during World War II, when welded structures were widely introduced into the shipbuilding industry. Numerous solutions improving the fracture toughness (inhomogeneous structures, multiphase, fine grain materials, etc.) were obtained with the help of fractography.

The above introductory remarks explain the importance of fractographic analysis. During the last few decades a lot of research work has been done in order to evaluate objectively fracture surfaces in a quantitative way.[8,15,25,30,41,42,49,72,78,88,92,101,106,107,112] Only a small part of this work is devoted to image analysis. The main reason lies in the fact that fracture surfaces are extremely difficult for automatic analysis, because many features are either impossible to clearly define and detect or researchers preferred the use of cheaper tools for data input, mainly tablet digitizers. However, currently computers have enough computing power and the latest versions of image analysis packages are relatively cheap. Therefore, image analysis should be much more widely used in fractography. In the next sections some problems as well as some possible solutions concerning the application of image analysis tools in quantitative fractography will be presented.

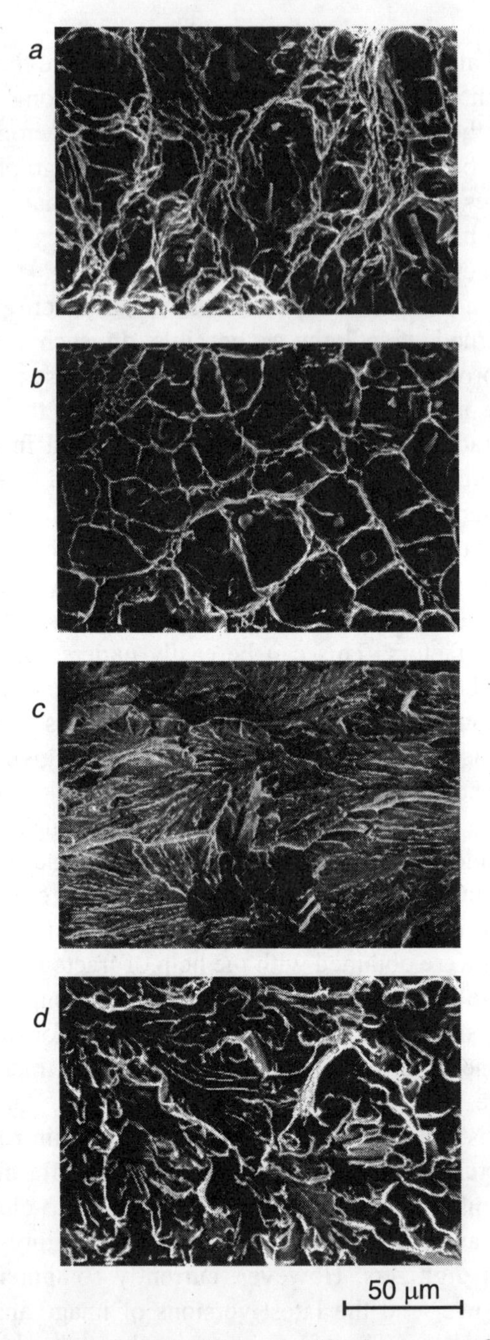

Fig. 5.8. Examples of fracture surfaces: ductile fracture surface of a low-carbon steel with numerous non-metallic inclusions (a), typical ductile fracture surface with dimples formed on precipitations (b), typical cleavage fracture surface with a river basin pattern resulted from mis-matching of the cleavage planes (c) and a fatigue fracture surface of a heavily overloaded construction (d).

The complexity of the fracture process mechanism significantly increases with the complexity of the material. Thus, it is sometimes almost impossible to quantitatively describe the fracture surface. This happens, for example, in the case of fiber-reinforced composites (see Fig. 5.9). One can discuss about this figure at least three entirely different types of sufaces:

- fracture surface of the matrix,
- fracture surface of the fibers,
- side-walls of the fibers, being a result of the pull-out process.

Each of these surfaces should be described separately and final analysis should take all of them into account. Even if we succeed in this description, it will be extremely difficult to find a correct correlation between the fracture surface phenomena and properties.

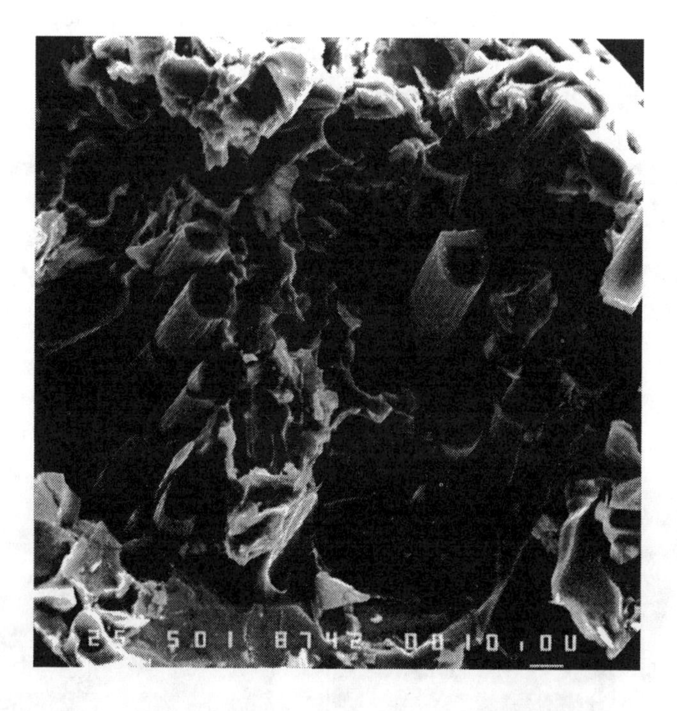

Fig. 5.9. An example of the extremely complex fracture surface of a fiber-reinforced composite.

Fracture surfaces can be divided into three basic groups:[33,107]

- *brittle fracture surfaces*, coming from fractures with low plastic deformation and built using relatively low energy. These surfaces usually contain almost perfectly flat elements, called *facets*, reflecting cleavage planes in grains as well as the river-basin pattern (Figs 5.8c and 5.10)

- *ductile fracture surfaces*, incorporating large plastic deformation and dissipating ten times more energy than brittle surfaces. Examples are shown in Figs 5.8a, b and 5.11
- *fatigue fracture surfaces*, formed as a consequence of time-dependent, usually periodical loads that are too low to form the immediate fracture of brittle, ductile or mixed character. An example is shown in Fig. 5.8d.

This division has some practical impact, as the type of fracture surface is closely bound with the amount of energy necessary to form the fracture surface. Therefore, it would be interesting to have tools for automatic recognition of the type of fracture surface being investigated. The most natural way to achieve this goal seems to be an analysis of the gray-level distribution. Unfortunately, even within a single fracture surface we can get a series of completely different distribution functions (Fig. 5.10) and very similar distributions for both brittle and ductile surfaces (Fig. 5.11).

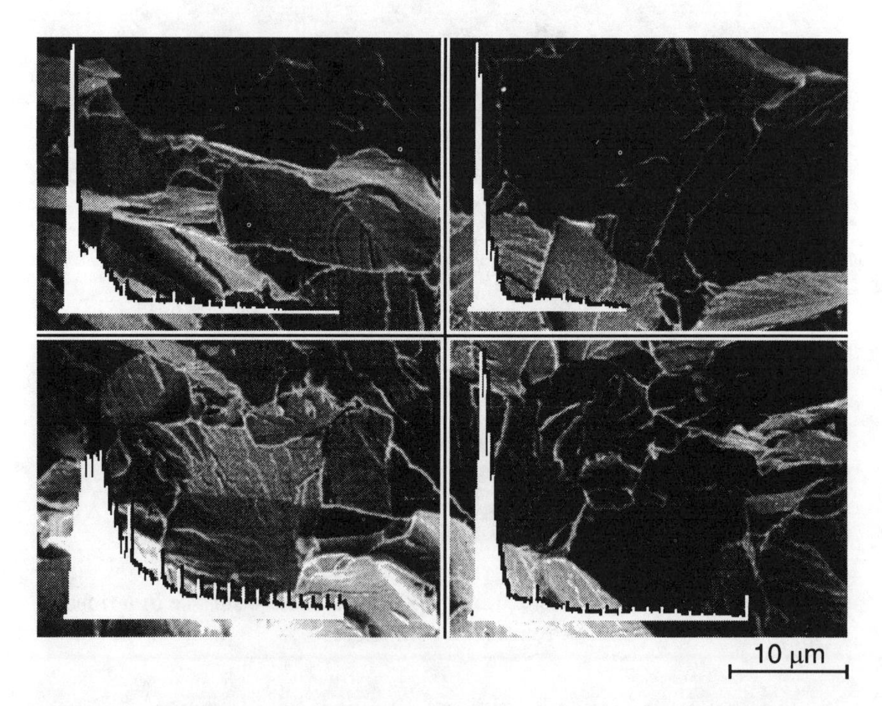

Fig. 5.10. A brittle fracture surface divided into four regions and the corresponding gray-level distributions shown in white. These distributions are clearly different.

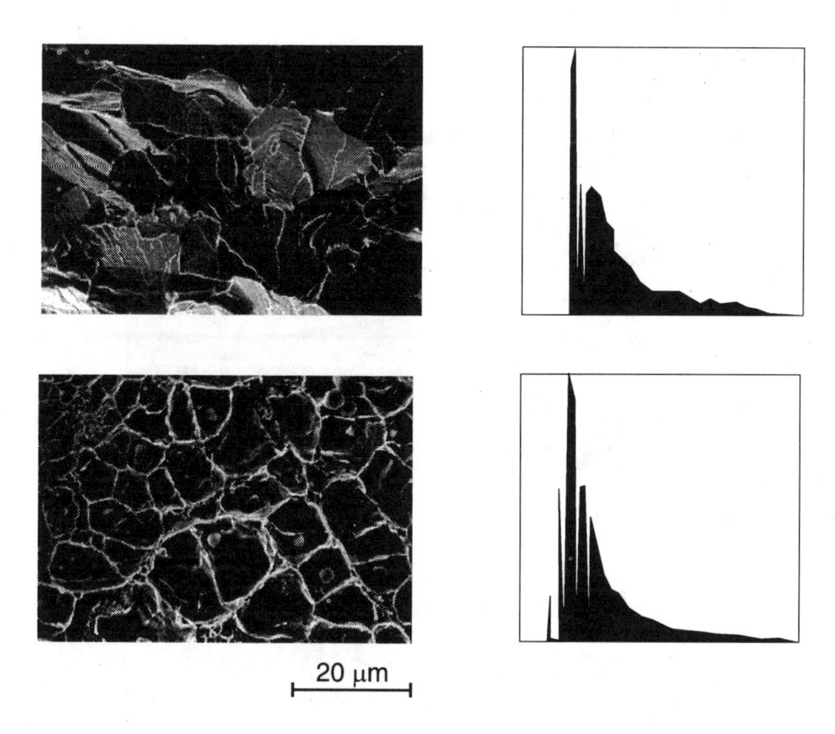

Fig. 5.11. Comparison of two, different fracture surfaces: brittle (top) and ductile (bottom) together with the corresponding gray-level distribution functions, which are very similar.

So, gray-level distribution is not a good tool for recognition of the type of fracture surface. Closer analysis[81,110] indicates that the character of the fracture surface can vary significantly with magnification. Surfaces looking ductile at low magnification can have a dominating brittle character, visible at higher magnification. Similarly, macroscopically brittle surfaces can, in fact, possess a ductile character which can be revealed only at high magnification. This observation explains why the division discussed above has less practical meaning than it should have from the theoretical point of view.

At the early stages of quantitative fractographic analysis Broek[8] showed that the toughness of a ductile steel is proportional to the distance between non-metallic inclusions. This observation was supported by much further research. An even more accurate correlation can be obtained if the inclusions or other particles are investigated directly on the fracture surface.[106] Such an analysis does not require specialized algorithms. It is sufficient to apply tools described in Section 4.3.

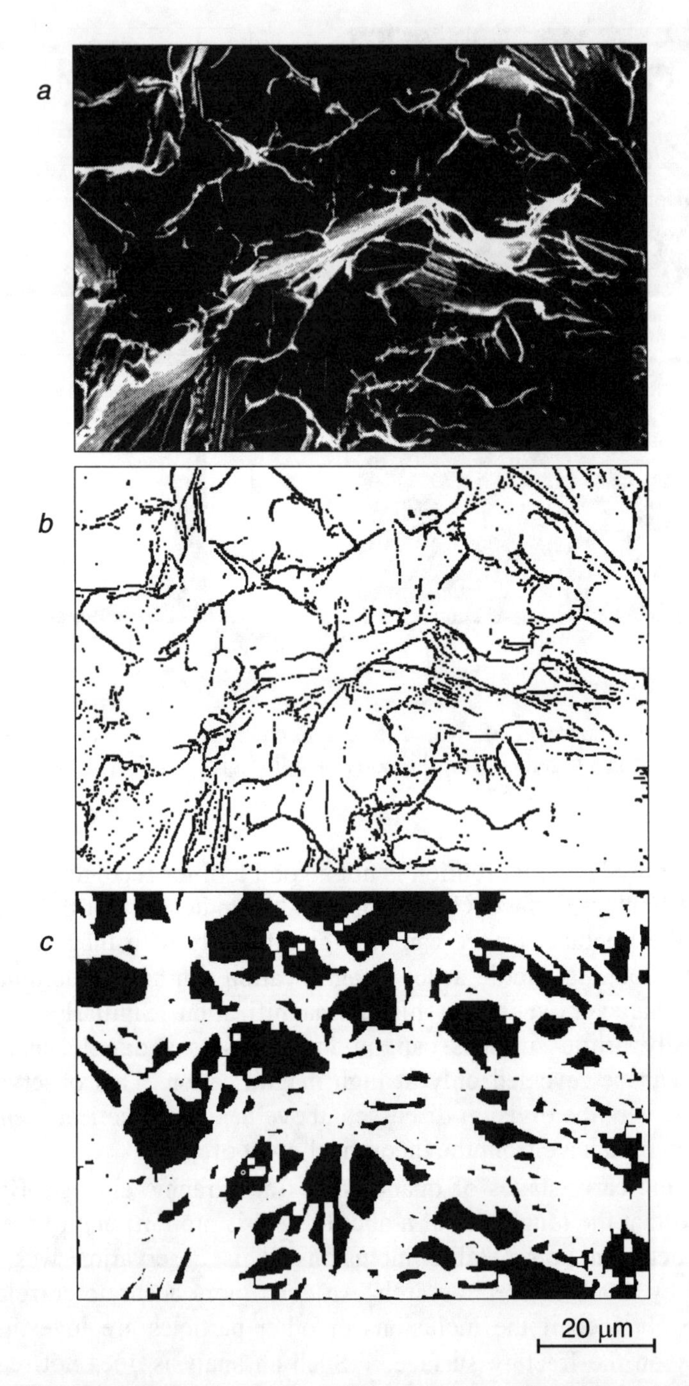

Fig. 5.12. Detection of facets in a brittle fracture surface. Initial image (a), detected edges (b) and detected facets (c).

Image analysis is a good tool for detection of the most characteristic features in fracture surfaces: facets in brittle one and dimples in ductile ones. An example of facet detection is shown in Fig. 5.12:

- facets in the initial image (Fig. 5.12a) are separated by more or less clearly visible edges. The first step in facet detection is to reveal the edges. This can be done using any suitable algorithm for edge detection. Numerous examples are presented in Chapter 4, the choice depends on the morphology of the analyzed surface. The detected edges are presented in Fig. 5.12b
- the final detection is now quite straightforward. Erosion produces continuous network of edges, then the size-dependent filtering or erosion with subsequent reconstruction (a series of dilations and subsequent AND with the initial image repeated until convergence) leads to detection of the main facets. The detected facets are shown in Fig. 5.12c as black spots (for a better visualization effect the final image was inverted).

A technique very similar to the algorithm discussed above can be applied for detection of dimples in ductile fracture surfaces. An algorithm developed for grain boundary detection, discussed in Section 4.1, example 8, can be successfully applied without any modification for these purposes.

Another feature which can be successfully detected using image analysis techniques is the stretch zone width (SZW) at the crack tip. It has been shown that during the fracture process of metallic materials a heavy plastic deformation occurs at the crack tip prior to crack growth. This deformation causes blunting of the crack tip and one of its possible measures is the crack opening displacement (COD), used since the sixties as a measure of fracture toughness. If we measure the same phenomenon in a perpendicular plane (it is just the fracture surface) we get a new measure, the above-mentioned SZW.[41,107,108] An international round-robin program[41] has shown that the scatter of results of manual SZW measurements can be compared with the measured value. Unfortunately, measurements using image analysis have not been applied in this program. However, it is evident that every algorithm will give the same results, irrespective of how many times it is applied to the same image. By contrast, in manual measurements we get more or less different results every time, even if the measurements are done by the same person. Moreover, for a practical application we do not need absolutely precise values. If there is a systematic, constant error the results can be successfully applied in practice. An example of SZW detection is shown in Fig. 5.13.

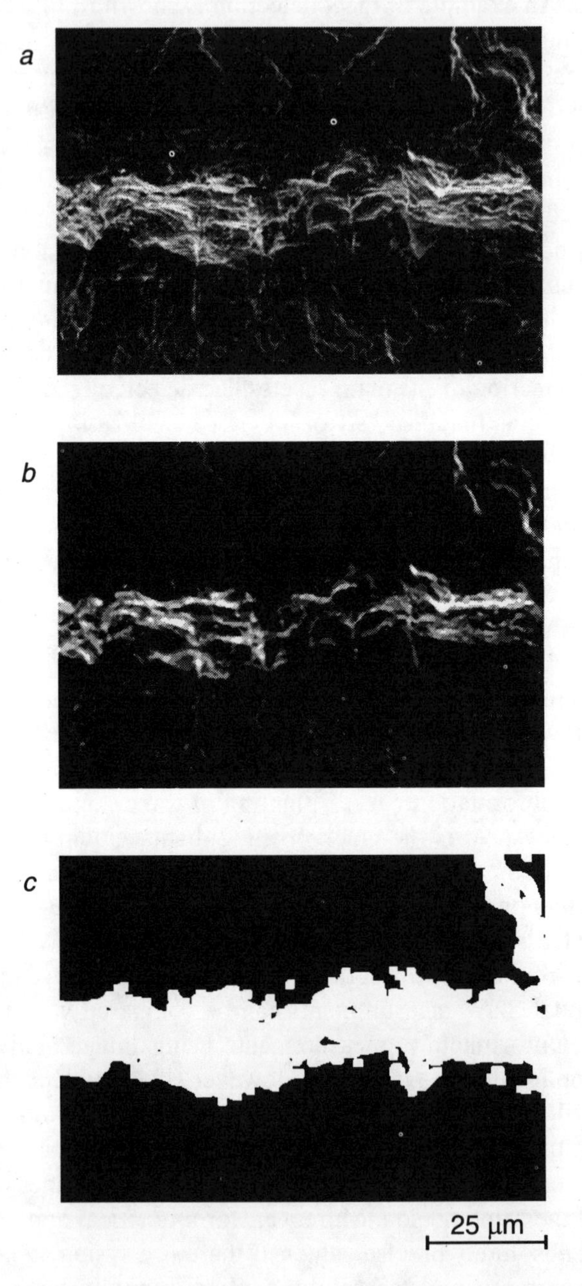

Fig. 5.13. Detection of the stretch zone width (SZW). Initial image (a), initial image subtracted from the closed image with the subsequent normalization (b) and final detection performed by binarization and dilation (c).

The above analysis and examples clearly show that quantitative fractography in general does not require any special algorithms for detection of chosen features in the fracture surface. Algorithms developed for grain boundary or isolated particle detection can be applied. The main difficulty lies in a proper choice of the features to be detected and further interpretation of the results. This text, however, is devoted only to image analysis techniques. Details concerning the fractographic analysis can be found in the broad literature devoted to this subject.[8,21,25,41,49,59,62,66,72,78,101,106-108,115]

Most of the quantitative fractographic analyses has been performed on fracture profile lines, obtained from sections perpendicular to the macroscopic fracture surface plane. The theoretical background of profilometric analysis is well elaborated and supported by numerous practical applications.[30,101,107,108,112,115] The use of profiles has a lot of advantages:

- in contrast to the fracture surface, its profile line is easy for any mathematical analysis - we can compute length, curvature, angular distribution, height distribution, fractal dimension, etc.
- analysis of profiles can be performed simultaneously with analysis of the underlying microstructure, thus enabling detection of any structure-fracture interference
- only profiles enable analysis of heavily curved fracture surfaces with overlaps or secondary sub-cracks
- profile lines can be relatively easily recorded using tablet digitizers. Obviously, having access to the image processing tools can be very helpful and this point will be briefly discussed below.

Profiles suitable for quantitative analysis have to be prepared without any deformation. Therefore the first step in specimen preparation should be covering the fracture surface with a protective layer, usually made using electrolytic deposition of nickel. This layer is visible in Fig. 5.14a as a bright ribbon. Depending on the material being investigated, the profile line is already visible in a polished section without any additional processing or etching. Good chemical resistance of the nickel layer prevents it from etching which leads to a better initial image, with a clearly separated protective layer and the underlying specimen (Fig. 5.14a). Now we will see how to extract the profile line from the initial image:

- the most natural way of line detection seems to be application of any filter for edge detection (Prewitt filter in the case of Fig. 5.14b). The most clearly visible edge is detected at the interface between the protective layer and mounting resin, visible as a continuous thin bright line

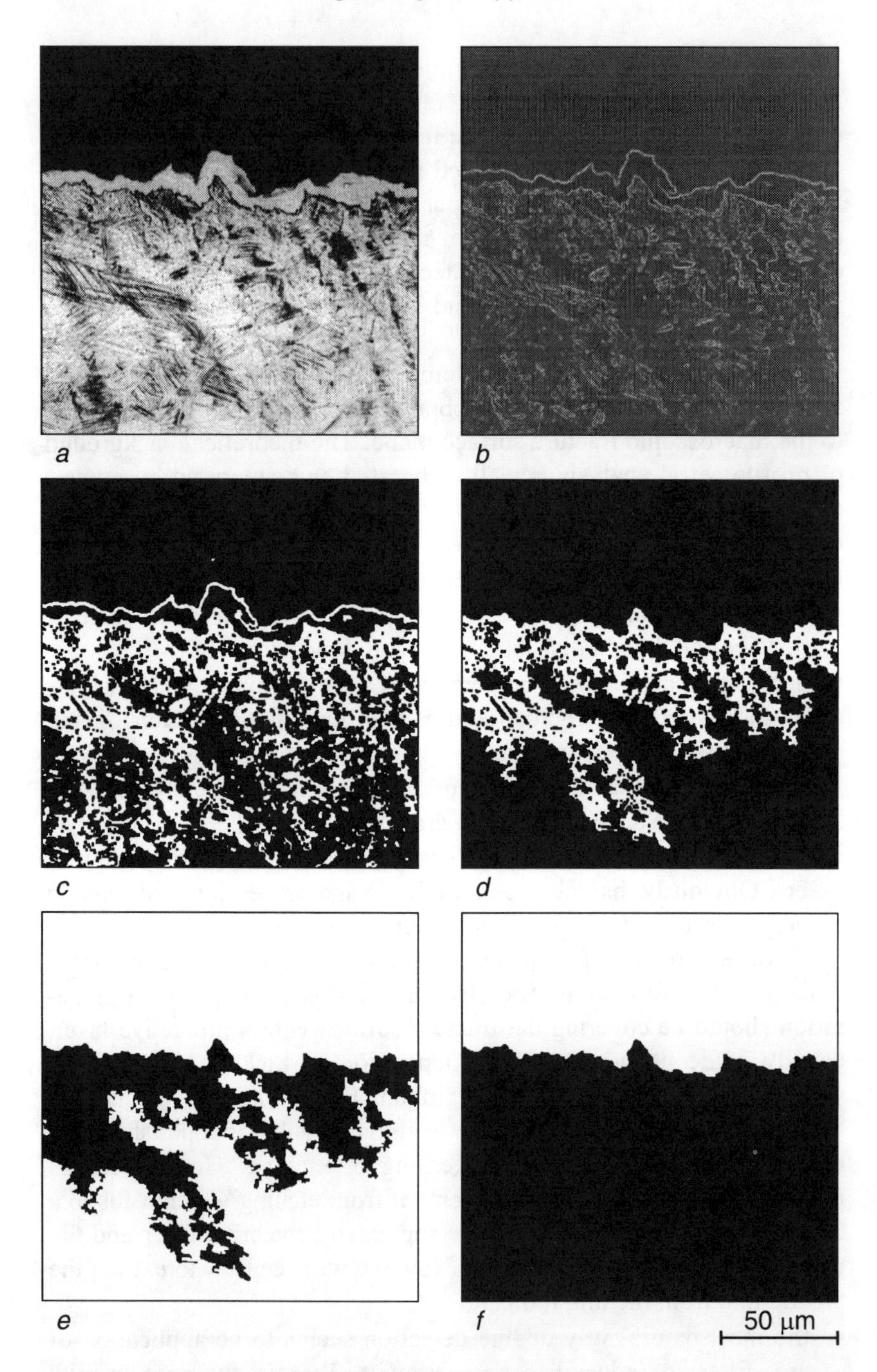

Fig. 5.14. Detection of the fracture profile. Initial image (a), edges produced by the Prewitt filter (b), binarized edge image (c), erosion (in order to remove small and thin objects) with subsequent reconstruction (d), hole filling and negative (e) final detection (f).

- binarization produces a clear image of the detected edges which is suitable for further processing (Fig. 5.14c)
- a small erosion will remove the unwanted pseudo-profile line at the top of the protective layer as well as small, noisy features
- reconstruction of the binarized image starting from the eroded set will give a better defined image of the edge (Fig. 5.14d)
- subsequent hole filling assures that the profile is continuous. The negative of the hole-filled image gives a huge white region at the top of the image (Fig. 5.14e). This can be easily detected, for example, using the labeling technique, which leads to the final image (Fig. 5.14f).

The profile line can be obtained from Fig. 5.14f without any difficulty, by erosion and a subsequent XOR operation with the image before erosion.

If the contrast between the protective layer and the rest of the specimen is high, detection of the profile line can be completed without edge-detecting filtering. This is presented in Fig. 5.15, using for better comparison the same initial image as in Fig. 5.14. If we threshold the initial image we will get a binary image (Fig. 5.15b) which differs significantly from the binary image obtained after edge-detecting filtration (Fig. 5.14c) and therefore requires a rather different treatment:

- we have a wide ribbon of the binarized protective layer which can be identified using the labeling technique (Fig. 5.15c) and finally detected by binarization of the labeled image (Fig. 5.15d)
- we have in Fig. 5.15d three objects: the black area of the bonding resin (top), the white protective layer and the black area of the specimen (bottom). After inverting the image we get two white regions, separated by a black belt. Applying labeling and binarization once more we get the final detection (Fig. 5.15e)
- multiplying images from Fig. 5.15a and e produces the mask image which we can check to see how satisfactory our detection is (Fig. 5.15f).

Having the detected profile line we can analyze it applying all the parameters used in quantitative fractography. This analysis can be effectively coupled with detection of any structural constituents, leading to such parameters as the fraction of a given phase in the formation of the fracture surface. This can clarify numerous questions connected with the fracture mechanism.

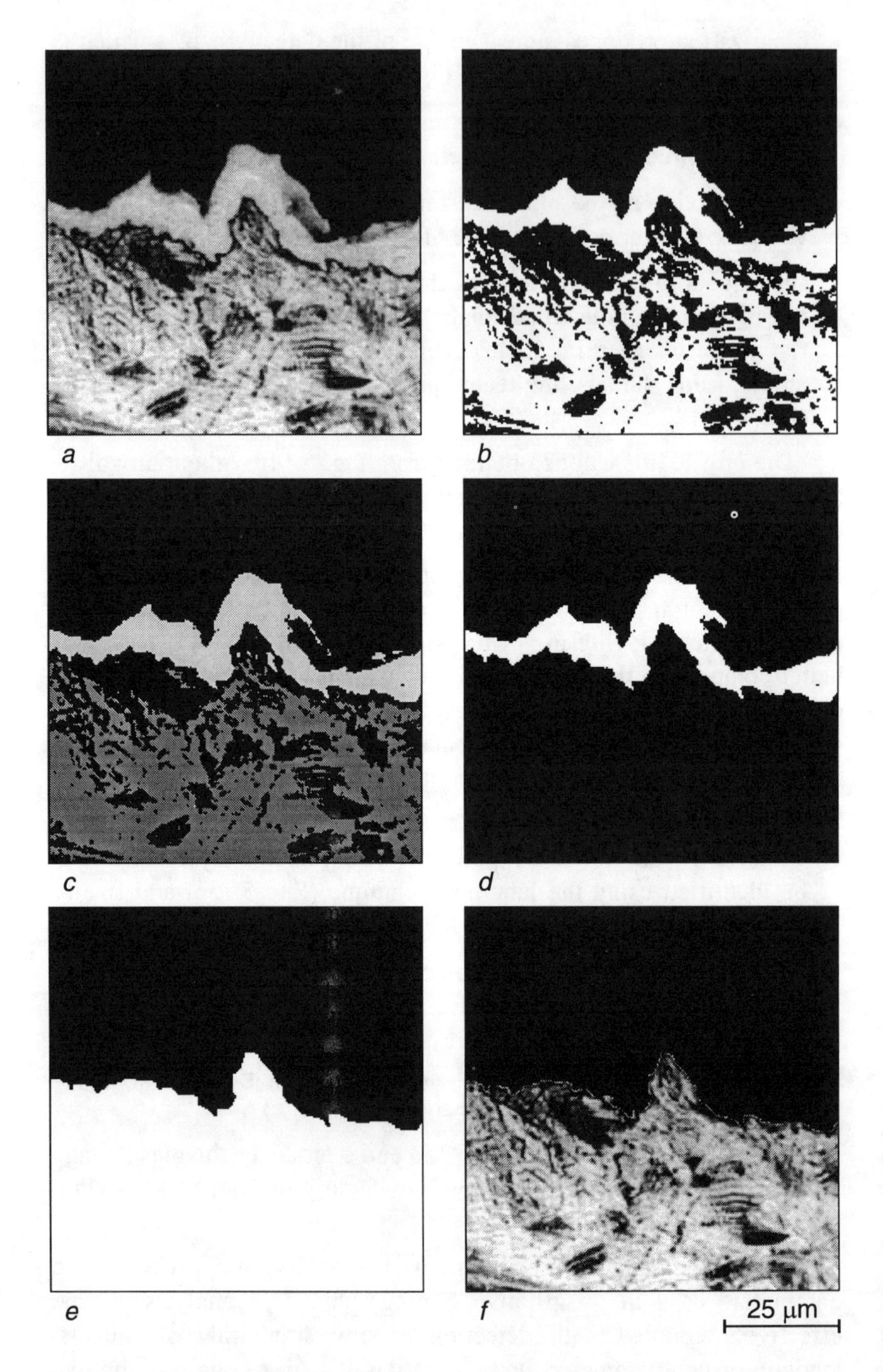

Fig. 5.15. Another way of profile line detection. Initial (a), binarized (b) and subsequently labeled (c) image, detected protective layer (d), final profile detection (e) and illustration of the detection quality (f).

Chapter six

Analysis and interpretation

In all the previous sections we discussed operations or functions which take images as arguments and return other images as results of transformations. If we try to describe these results in a formal way, we will get only a *qualitative* characteristic. In this chapter we will analyze another family of procedures, giving as results various sets of numbers, suitable for a *quantitative* description of the image.

6.1 Image processing and image analysis

Image processing and image analysis are often confused or used interchangeably. In this text we understand these terms as two bound together but clearly distinct processes. We will discuss this below in more detail.

Image processing is a sequence of several operations on a single image or a series of images, leading to a new set or single image. In other words, we manipulate the image in order to suppress the waste information, extract necessary elements, improve the quality or readability of the image or simply achieve a spectacular artistic impression. Thus, image processing is widely applied in a wide variety of human activities, from leisure time to military operations. Image processing can be done using light and chemical processes in a darkroom, but in most cases is performed using computers and appropriate software.

Image analysis is a process of image data transformation, leading to some actions, decisions or conclusions. In this sense all of us, except blind people, perform continuous image analysis. A chemist in a pharmacy analyzes usually almost unreadable hand-written text and gives us the correct drug. A driver analyzes the scene in front of the windscreen and decides how to move the steering wheel. A customer in a shop analyzes the shape and graphics of packaging before making the decision of what to buy... etc.

Generally speaking, people possess an excellent ability to recognize or rather *interpret* images - we can easily decide what the contents of the image are. This corresponds with poor results to any *quantification* of the image. We cannot judge quickly and precisely how many windows are in a sky-scraper, what element is lacking in a complicated circuit, which seeds are too small or too big for automatic processing, etc.

The need for quantification is nothing new. However, for centuries it was enough to *quantify in a qualitative (!) way*: larger-smaller, higher-lower, longer-shorter, etc. The need to supply precise numbers is relatively new. It was born with the contemporary digital technique that uses bar codes and computerized production processes.

Simultaneously, stereological methods (stereology was formally born in the sixties) have undoubtedly proven that there exist precise, fully quantitative relations between the materials microstructure and properties.[33,34,50,66] Obviously, in order to establish such relations, microstructure has to be described by numbers. Coming back once more to the above-mentioned *qualitative quantification*, in the case of materials it is enough to quantify the following categories:

- *amount* of all the structural constituents and (separately for each constituent),
- *size* of the particles or their colonies,
- *shape* of the particles or their colonies and
- form of their *distribution* over the material volume.

The idea is very simple but we have no good measures suitable for quantification of shape or distribution. Therefore, finding a way to adequately quantify is still a real problem. Some solutions can be found in the literature devoted to stereology.[28, 54,79,82,100,104]

Stereological methods constitute an efficient set of tools for microstructural quantification. However, in order to obtain results with sufficient accuracy it is necessary to repeat simple measurements on hundreds of particles or other features. This is a very tiring and time-consuming process of repeated operations, thus an ideal goal for computerization. In this way we come to the conclusion that computers should be applied for any measurements and further statistical analysis of the results. This analysis is sometimes called *image understanding*. The whole process (starting from the initial image) is called *computer-aided image analysis*. Obviously, any images for such treatment should be appropriately prepared, *using image processing*. To summarize, we can state that image processing is a part of the image analysis process (Fig. 6.1).

In Fig. 6.1 one can see three ellipses, symbolizing image analysis, processing and understanding, respectively. These symbolic ellipses are bound together in a quite complicated way. This illustrates well the complexity of relations among the components discussed:

- image processing is a part of image analysis and that seems to be clear. Another part of image processing is situated outside image processing because some artistic filters (for example, converting the initial image into an impressionistic style painting) are not used in image analysis
- image understanding is fully included in image analysis and has a small commonality with image processing as some processing tools are used for measurements (for example, logical operations)
- some portion in the image analysis ellipse is not occupied by other ellipses. This represents unique, advanced image analysis operations, like, for example, the use of expert systems, neural networks etc.

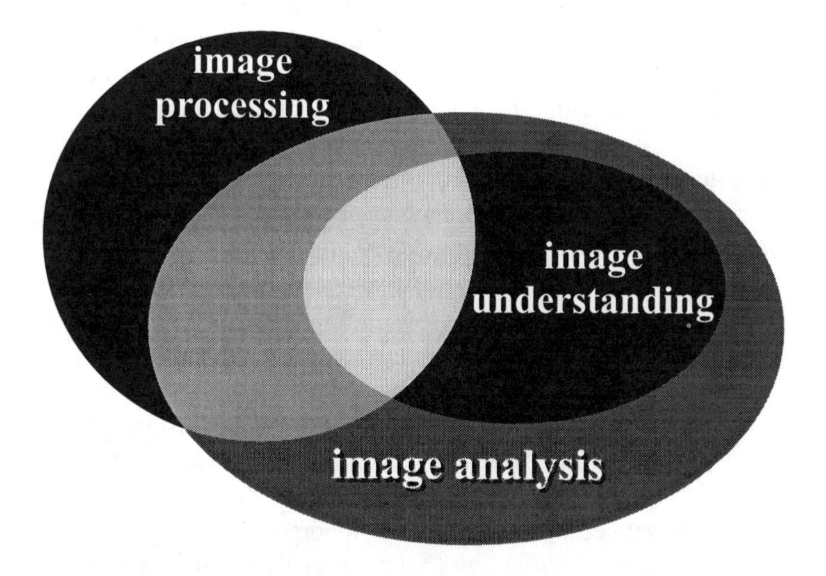

Fig. 6.1. Illustration of the relationship among image analysis, image processing and image understanding.

6.2 *Measurements of single particles*

Prior to any measurements of any particle the program has to unequivocally decide where the edges of this particle are. For us this is a quite straightforward operation. Unfortunately, for a computer, this is not so simple. In practice we need to have a binary image for this purpose. This explains why so many algorithms, presented in the previous sections, lead to binary images as the final result of analysis.

In this text we avoid theoretical considerations and focus on applications. Consequently, we will skip all the data concerning the nature and methodology of digital measurements, describing only the chosen measures and their possible application. We will start our discussion with an analysis of measurements of a single particle. The basic quantities are schematically illustrated in Fig. 6.2:

- in Fig. 6.2a we have the initial particle. This particle is inhomogeneous (this property will be discussed later) and for further analysis should be converted into a binary form (Fig. 6.2b)
- probably the most natural measure which can be evaluated using computerized tools is the particle surface *area*. In a binary image (Fig. 6.2b) area can be easily measured by simply counting the pixels forming the particle
- the next measure, frequently used in quantitative analysis is the particle *perimeter* (Fig. 6.2c). The digital, discontinuous nature of computer images means that the perimeter is usually computed with large errors and this should be taken into account by the users. Some solutions suitable for improvement of the accuracy of this measure will be discussed later in this section
- very popular, fast and accurate measures are the so-called *Feret diameters*, characterizing the outer dimension of the particle. In Fig. 6.2d we have two, horizontal and vertical Feret diameters
- more advanced packages allow measurement of a *user-oriented Feret diameter* (Fig. 6.2e). In this case we get two numbers, describing the length and orientation angle
- simple doubling of the maximum value of the distance function gives an interesting measure, namely, the *maximum particle width* (Fig. 6.2f)
- among advanced measures a very important one is (Fig. 6.2g) the *maximum particle intercept* (not necessarily equal to the maximum Feret diameter!). This measure is also characterized by two numbers, describing the length and orientation angle
- for some applications the *coordinates of the center of gravity* (Fig. 6.2h) are very important
- a similar value has the next measure, *coordinates of the first point* (Fig.6.2i). First point is chosen as the most left-situated pixel from the top row of particle pixels
- in the case of concave particles interesting information can be obtained from the *convex hull* (Fig. 6.2j). All the measures presented can be applied to the convex hull as well as to the initial particle
- rather similar to the convex hull is the *bounding rectangle* (Fig.6.2k), suitable for shape characterization

- the last quantity discussed here is the *number of holes* (Fig.6.2l).

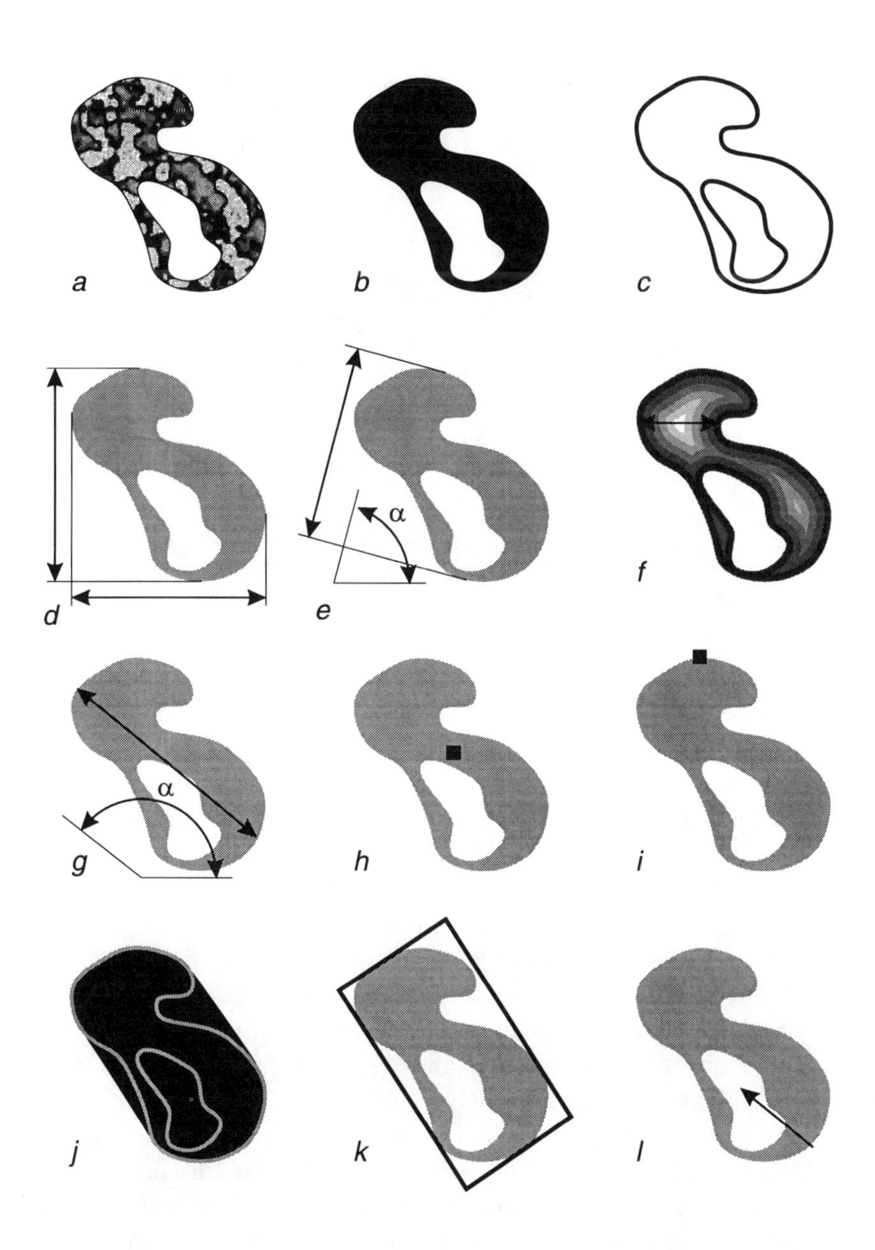

Fig. 6.2. Basic measures of a single particle: initial particle (a), area (b), perimeter (c), Feret diameters (d, e), maximum width (f) and intercept (g), coordinates of the center of gravity (h) and the first point (i), convex hull (j), bounding rectangle (k) and number of holes (l).

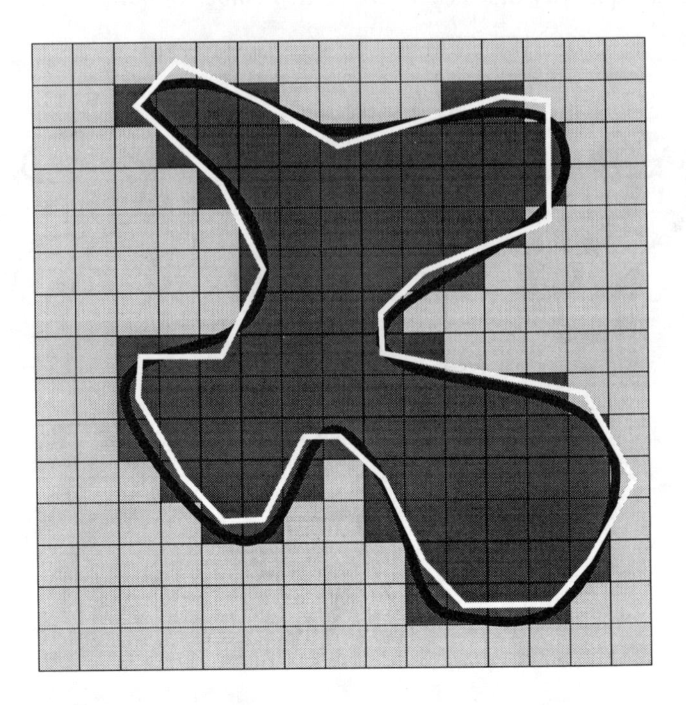

Fig. 6.3. A particle (black line) together with its pixel (gray squares) and vector (white line) representation.

Binary images devoted to measurements should be very carefully prepared. The image analysis program will perform the measurements on any image, without taking into account its quality. Therefore, the user should keep in mind some general safety rules:

- decide what rule of connectivity (see Section 4.1, Fig. 4.15) will be applied. If we apply the wrong rule, some particles can be over-segmented or others can be erroneously glued together
- check the size of the smallest particles. Generally, if any particle is made up of less than approximately 10 pixels, the results of the measurements are extremely inaccurate
- check in the documentation how your software computes the perimeter. This is often the main source of errors, as perimeter is used for evaluation of numerous stereological parameters, especially shape factors. Two different packages can give identical values for almost all the parameters while the difference in the perimeter can exceed 50%.

The remarks above suggest that one should pay special attention to the results of perimeter evaluation. Let us analyze now the reason for these problems. A particle usually has a smooth edge (black line in Fig. 6.3) which has to be approximated using the pixel grid, possessing obviously limited resolution (see Fig. 6.3; small squares symbolize single pixels). It is obvious that simply counting the pixels forming the particle boundary is insufficient. It is always possible to compute the perimeter as the weighted sum of these pixels with weights being a function of a local pixel configuration. Unfortunately, this solution also does not prevent possibly large errors.[114] A much better solution is to create the vector representation (also called the polygonal approximation) of a particle - an effect very often used in computer graphics. This vector representation consists of a series of nodes and lines connecting them (see the broken, white line in Fig. 6.3). The length of this line can be exactly computed and provides a good approximation of the real perimeter. In the most advanced version the nodes can be connected by curved sections, called splines, giving an even more precise approximation.

The last, discussed here, method of rough but very quick approximation of the perimeter, is an application of the method known as a Crofton formula.[114] Here we will make an exception to the general rule that we do not introduce theoretical considerations. The reason is that in some cases the use of the Crofton formula can be really helpful and appropriate computations can be easily performed using the macros built-in into every more advanced image analysis package.

The Crofton formula is based on an analysis of *the total projected lengths.*[82,100] The projected length at an angle α, $D(\alpha)$ (Fig. 6.4a) is identical with the user-oriented Feret diameter (Fig. 6.2e). A new measure, however, is the total projected length (Fig. 6.4b). For convex particles both quantities, projected and total projected lengths, are identical. In the case of concavity the total projected length is a sum of partial projections, as shown in Fig. 6.4b:

$$D(\alpha) = D_1 + D_2 \tag{6.1}$$

This can be easily and exactly computed in digital space. The pixel pattern in the digital image forms a series of lines, with the inter-line spacing *dx*. In order to compute the total projected length it is enough to count the number of pixels representing the *line-entry-points* (denoted by small, black circles in Fig. 6.4c) and multiply this number by the *dx* spacing. The line-entry-points can be successfully detected for different orientations using the HMT transformation and the following structuring elements:

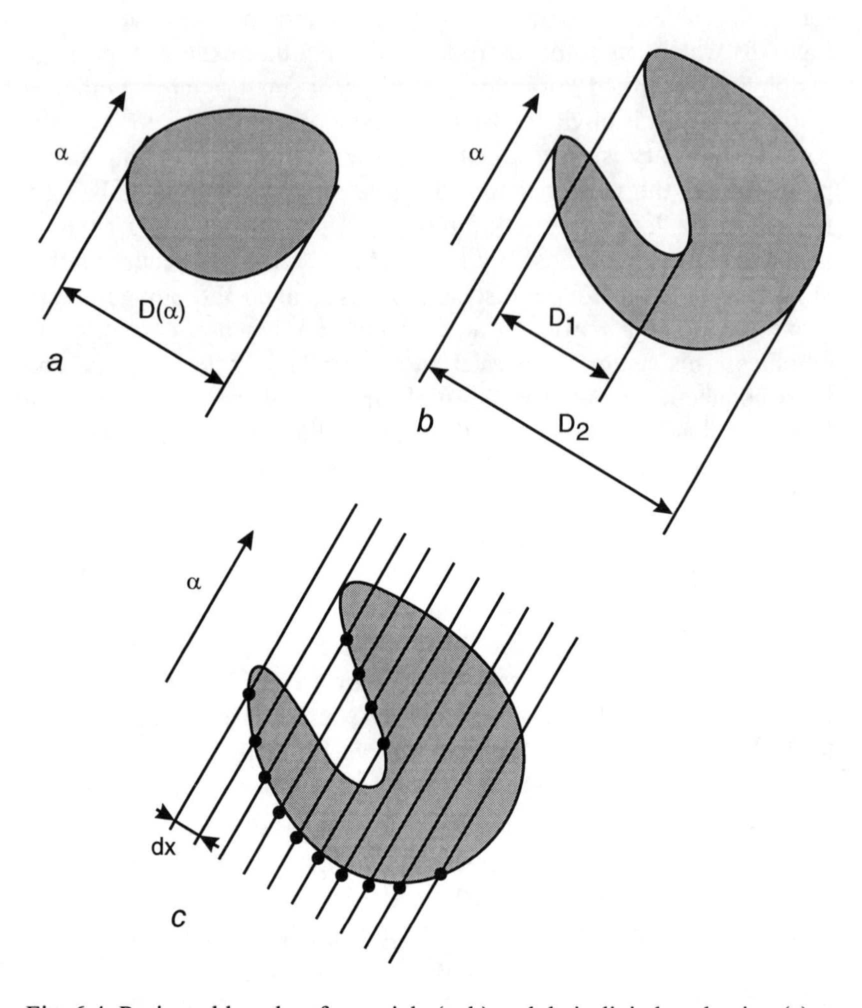

Fig. 6.4. Projected lengths of a particle (a, b) and their digital evaluation (c).

for orientation angle 0°

X	X	X
X	0	1
X	X	X

for orientation angle 45°

X	X	1
X	0	X
X	X	X

for orientation angle 90°

X	1	X
X	0	X
X	X	X

for orientation angle 135°

1	X	X
X	0	X
X	X	X

Cauchy has introduced the following rule, allowing evaluation of the perimeter from total projected lengths:

$$L = \int_0^{\pi} D(\alpha) d\alpha \tag{6.2}$$

where: L is the perimeter, α is the orientation angle and D(α) is the projected length.

Let us analyze eqn. (6.2) for a circle. Its total projected length is identical for all orientations and equal to its diameter *d*. After integration we get the well-known formula for evaluation of the perimeter of a circle: $L=\pi d$.

In the case of an hexagonal grid it is enough to make measurements in three basic directions and we get the following final formula:

$$L = \frac{\pi}{3} \times \frac{a\sqrt{3}}{2} \times \left(N_0 + N_{60} + N_{120}\right) \tag{6.3}$$

where: a is the distance between two neighboring pixels, and N_0, N_{60} and N_{120} are the numbers of line-entry points for 0, 60 and 120 degrees, respectively.

In an analogous operation for the square grid one has to take into account the difference between the neighboring pixel distance for angles 0 and 45 degrees. The appropriate formula for perimeter evaluation will take the following form:

$$L = \frac{\pi}{4} \times \left[a \times \left(N_0 + N_{90}\right) + \frac{a}{\sqrt{2}} \times \left(N_{45} + N_{135}\right) \right] \tag{6.4}$$

where: a is the distance between two neighboring pixels, and N_0, N_{45}, N_{90} and N_{135} are the numbers of line-entry points for 0, 60 and 120 degrees, respectively.

The method of perimeter quantification described above sometimes gives due measurements limited to three or four directions, considerable errors. However, the method has no systematic, methodological errors and constitutes a very fast, efficient computational tool.

6.3 First measurements - numbers

The simplest and most natural type of measurement is simply counting objects. In every day life this usually does not require any additional tools, knowledge or skills except counting. If you need to buy some fruit for a meeting, you usually order one pound but you can order six pieces as well. If you want something to drink, you can order a liter or a single glass, etc.

Counting also constitutes the basic group of measurements in computer-aided image analysis. Counting objects has a lot of applications in further analysis. Some examples are listed below:

- if we divide the image surface area by the number of grains in a single-phase material we will get the mean section area, being one of the most commonly used measures of a grain size
- the number of particles per unit area is one of the important stereological parameters, characterizing the particle density
- the number of particles per unit area is used for estimating the number of particles per unit volume or the surface curvature
- counting the number of points having a defined neighborhood can be used for computation of the projected lengths (see previous section) or other parameters, not discussed in this text as we wish to avoid theoretical considerations.

Counting objects is done on binary images, usually using the labeling technique. *Labels* are numbers coupled with all the pixels in a binary image. The pixels in a matrix get the value of zero and the pixels in particles get ascending natural numbers in the following way: all the pixels in the first particle get 1, all the pixels in the second particle get 2, all the pixels in the third particle get 3, etc. The image is scanned line by line from left to right. If we hit a point of a particle not labeled yet, we give the next number to this pixel and extend this number to all the pixels of the same particle. Obviously, the rules of connectivity play an important role in this process. If we want to count objects in an image we simply need to find a maximum value in the labeled image and this maximum is directly equal to the number of particles.

So, counting objects is really very simple. Unfortunately, there is a risk of serious errors. Usually, some particles are crossed by the edge of the image. If we analyze the next field, touching the currently analyzed image, the same particles crossed by the edge will be partially visible in the next image and, consequently, counted twice. Now we will show how to avoid this systematic error.

Most of the image analysis programs allow automatic removal of the particles crossed by the edge of the image, often called *border kill*. There is a temptation to use this transformation in order to solve our problem, but closer analysis proves that this is a wrong solution. The reasons are twofold:

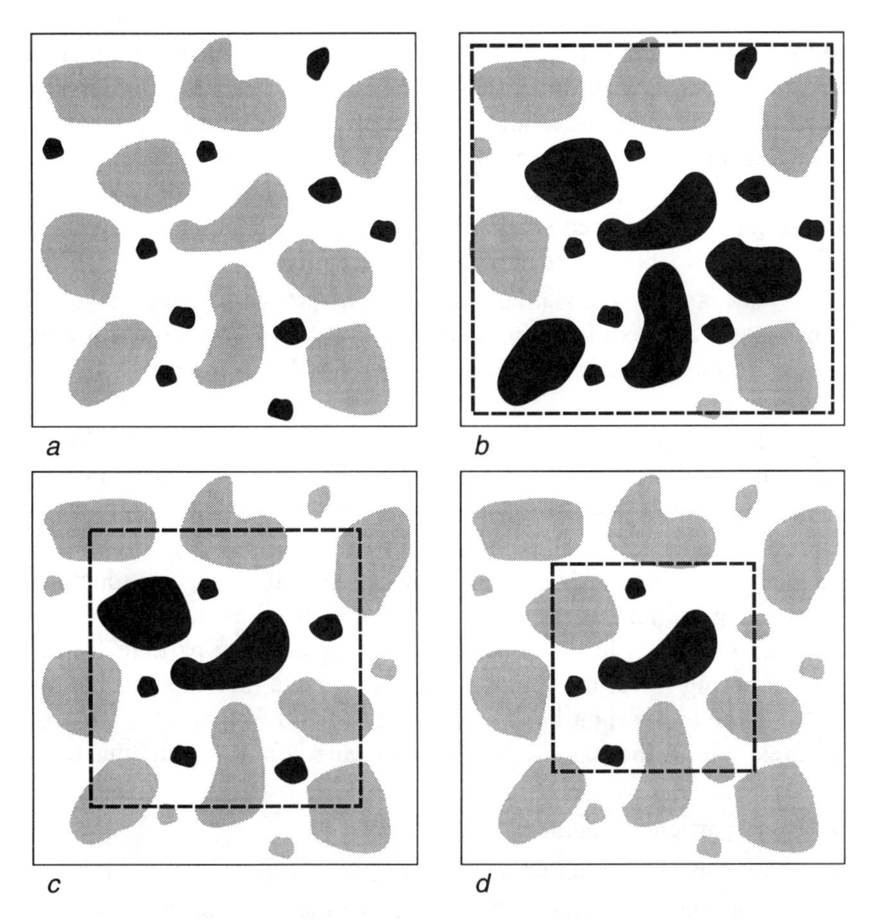

Fig. 6.5. Effect of the border kill operation on the distribution of the detected particles.

- first, instead of counting some particles twice we remove them and these particles are not counted at all. This underestimates the number of particles
- second, a border kill operation affects the distribution of particles.

Now we will discuss in more detail the second item: affecting the particle distribution. In some cases we are not interested in the exact number of particles, but rather in the distribution of some attributes of these particles, for example, size. The border kill operation is used in order to remove the particles which are crossed by the image edge and therefore cannot be correctly analyzed - if we have only a part of a particle we cannot measure its size. Unfortunately, this can heavily distort the examined distribution. To study this mechanism we will analyze Fig. 6.5. In Fig. 6.5a we have a test image with 10 large (gray) and 10 small particles (black), uniformly distributed over the image. 50% of particles are large. If we simulate taking a little bit smaller image (a square with broken sides in Fig. 6.5b) and the border kill operation, we will remove all the gray particles. Within the remaining particles, marked by black color, 8 are small and only 5 are large. So, in this new image only 38% of the particles are large. We can repeat this simulation for even smaller images (see Fig. 6.5c and d), getting 29% and 25% of the large particles, respectively.

This example demonstrates well that the probability of removal of a particle is approximately proportional to its size. Therefore, for correct counting of particles we have to use another technique, called a *guard frame*.

Let us analyze a set of particles, as shown in Fig. 6.6a. We introduce the guard frame, a rectangular area, shown in light gray in Fig. 6.6b. For an unbiased particle count we take into consideration all the particles totally included in the guard frame and touching or crossing its right and bottom edges. Simultaneously, all the particles touching or crossing the left and upper frame edge, as well as the upper right-hand or lower left-hand corners, are removed. If any particle fulfills two opposing rules, the removal rule is decisive (see the particle located at the upper right-hand corner of the guard frame). This concept assures that in the case of a series of guard frames touching each other:

- every particle is counted
- every particle is counted only once.

In Fig. 6.6c we can observe a result of the selection based on the concept of the guard frame (only particles drawn in black will be analyzed). This figure shows also that the rule described, although working well in the absolute majority of cases, can also produce overdetection. The particle indicated by an arrow in Fig. 6.6c will be counted twice because it crosses the lower edge in one frame and the right edge in another one. In order to prevent such (rare) cases we have to improve the classification rule. This can be easily done by

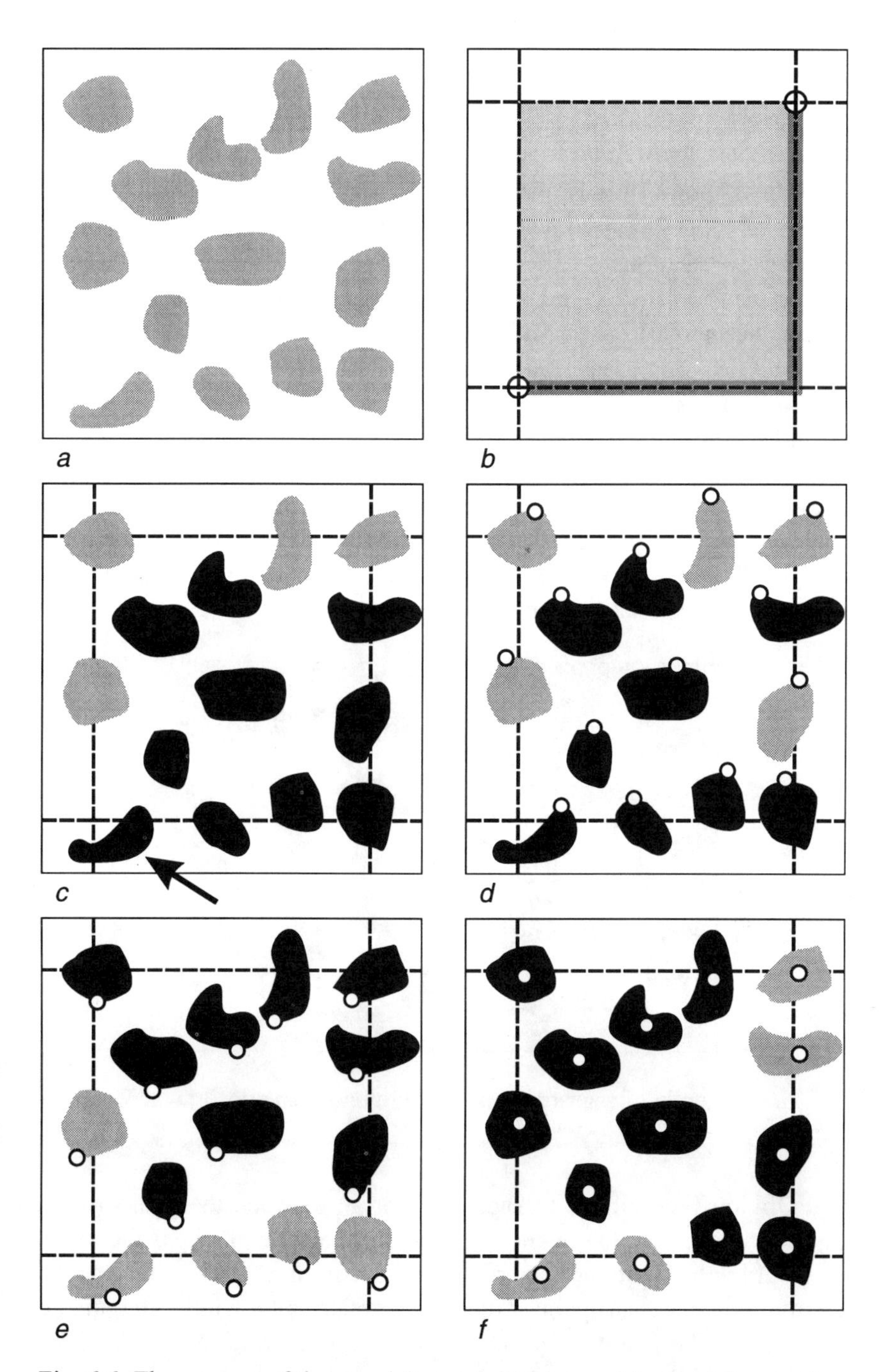

Fig. 6.6. The concept of the guard frame. Initial image (a), guard frame schematically (b), selection based on crossing the edge of the frame by the particles (c), selection based on location of the characteristic points: first point (d), last point (e) and center of gravity (f).

coupling the particle with a unique pixel defined in such a way that every particle should possess only a single pixel matching this definition. There are many possible definitions of such identification points. Here we will analyze three of them:

- the first point (this is a first point of a particle hit during scanning of the image line-by-line from left to right. The result of this classification rule is shown in Fig. 6.6d
- the last point (detected in a way analogous to the first point, described above). The result of this classification rule is shown in Fig. 6.6e
- the center of gravity. The result of this classification rule is shown in Fig. 6.6f.

Comparison of the results of classification shown in Figs. 6.6 c, d, e and f demonstrates that every rule produces different results. This is true, but on the other hand, every particle is taken into consideration (only within another frame) and is counted only once. Thus, the method described above can assure unbiased counting.

Fig. 6.7. Example of a guard frame placed non-symmetrically within an image.

Obviously, we have to keep the margin between the guard frame and the image edge large enough to analyze all the pixels of the particles cut by the right and lower edges of the frame. So, the guard frame should not necessarily be situated symmetrically within the image (Fig. 6.7). The guard frame is drawn here as a white rectangle. This frame should be kept far enough from the left and bottom edges of the image in order to keep the white-drawn particles out of the frame. These particles are cut by the image edges and therefore only a part of them is placed within the image. Simultaneously, as all the particles cut by the upper and left side of the frame are thrown away, no margin between the frame and image edges is necessary at this location. Try to guess which of the rules illustrated in Fig. 6.6 was applied in Fig. 6.7. The correct answer is: the first point.

6.4 *Shape*

Shape is a very important factor in both qualitative and quantitative analysis of the microstructure of materials. This property can affect a lot of properties:

- shape (and orientation) of the grains in a single-phase material controls the anisotropy of mechanical properties
- shape of cells in polystyrene foam is closely connected to its insulation properties
- shape of the aggregates used for sintering can heavily affect the final density and microstructure of sinters
- shape of the graphite precipitates in cast iron affects so clearly the properties of this material that cast iron is classified on the basis of the shape of graphite precipitates (gray, vermicular or nodular iron)
- elongated non-metallic inclusions can cause embrittlement and delamination of steel plates.

The above list is not exhaustive, it contains only some selected examples illustrating the validity of shape in microstructural analysis. Shape is simultaneously very important and extremely difficult to evaluate. The primary reason for these difficulties is the lack of precise definition of the shape. In numerous books devoted to quantitative analysis of structure we can read that the shape of the particles is characterized by a set of shape factors, etc. No single word about the essence of this problem - what is shape?

The problem of shape definition is illustrated in Fig. 6.8a. In this figure we have three objects: a small circle, a square of medium size and a large star. We can separate these objects using shape criteria (still unknown to the reader) as well as the size criteria, for example, area or Feret diameter. Our everyday experience confirms that shape often changes simultaneously with size. For example, small and big cars have different shapes, the same happens with buildings, animals, plants, etc. Therefore, it is difficult to separate these two attributes. Moreover, shape is independent of size. Unfortunately, in the case of digital images, having limited resolution, this is not always the case. For example, in Fig. 6.8b we have a series of three figures (square, circle and star) grouped together. The difference in shape is evident for us if only the size (!) of the figure is large enough. The smallest objects in Fig. 6.8b cannot be separated using the shape criteria. Due to the approximation errors all these objects look similar. By the way, this explains the remark made earlier in this chapter, that we should avoid analysis of particles smaller than 10 pixels.

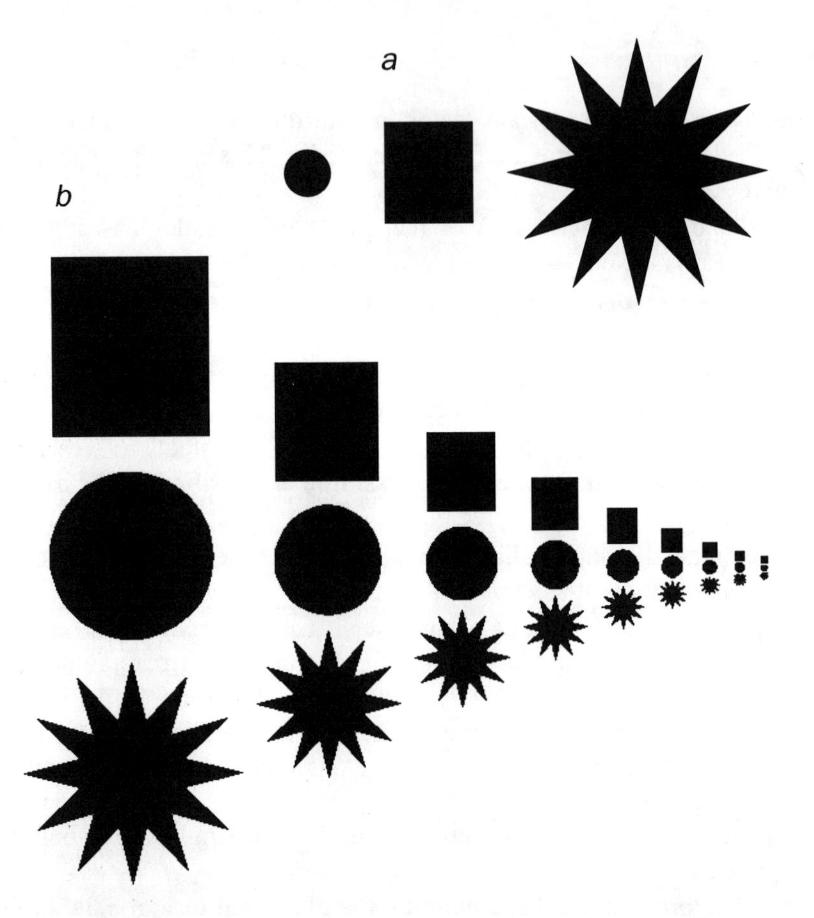

Fig. 6.8. Problems with shape definition. Quite often figures of different shape also have different size (a) and small particles look similar in spite of the differences in shape (b).

So, we know something about the difficulties with shape characterization but we still have no shape definition. However, we can try to define the shape of a flat figure in an indirect way, described below.

Let us assume that we have a series of figures with exactly the same surface area (Fig. 6.9a). These figures can be drawn in any way, with only one restriction: no figure can be formed from another one by translation, rotation and axial or central symmetry (Fig. 6.9b). We can now measure various attributes of these figures: surface area, perimeter, number of holes, curvature, etc. If the analyzed set is large enough, we always will find two figures with identical attributes. It can be, for example, identical perimeter or identical number of holes. So, none of these attributes can be used for an unique classification. The only attribute that differentiates all these figures is *shape*. In other words, shape is the way in which the figure can be drawn. For example, the following definition: *take a single central point and draw all the points equally distant from this central one* describes the shape of a circle.

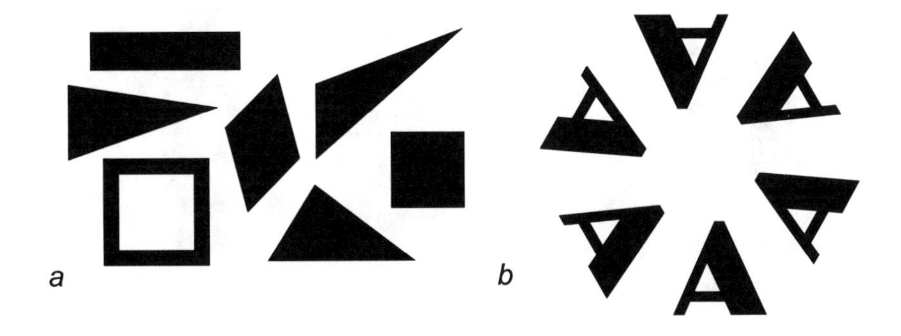

Fig. 6.9. A series of figures with various (a) and the same (b) shape.

So, shape is a very complex attribute. If you take into account that we can have an unlimited number of figures with different shapes, it becomes clear that shape cannot be (even theoretically) defined by a single number. The only thing we can introduce is definitions of various shape-dependent parameters which help us in quantification of the figures. A good example is the Cx coefficient which is used for characterization of the shape of cars. Knowing only this coefficient we cannot judge if the car is beautiful but this gives us *quantitative* information about how optimal its shape is from the aerodynamics point of view. A similar role in microstructural analysis is played by the so-called *shape coefficients* or *shape factors*.

Shape factors constitute a group of measures of a single particle or figure. They should be invariant to rotation, reflection or scaling. In general, shape factors are dimensionless (this assures that they are not sensitive to scaling) combinations of various measures. Below we will show simple but impressive examples of shape changes and corresponding shape factors. This will show the reader how these factors can be built within the image analysis programs. A more advanced study of this problem can be found in the literature devoted to stereology.

If we need to model any particle, the most simple and natural model seems to be a sphere. In flat projections (as images are) we will observe such ideal particles as circles. Therefore, one of the most frequent application of shape factor is testing how much a given shape differs from the circle. The analyzed shapes can be very far from the circle (see the right column of particles in Fig. 6.10) but the transition from the initial circle can be absolutely logical (analyze rows in Fig. 6.10).

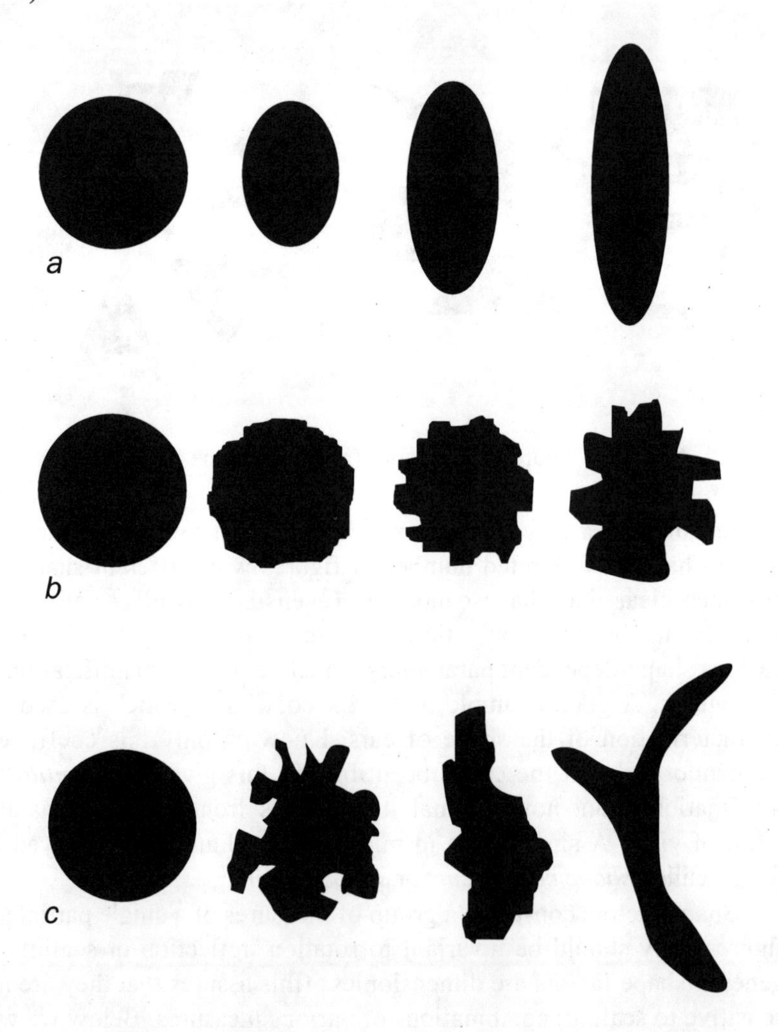

Fig. 6.10. Three families of shapes originating from a circle: ellipses of various elongation (a), shapes with various irregularity of the edge (b) and a combination of the above tendencies (c).

The three cases illustrated in Fig. 6.10 can be recognized for the needs of better text clarity as:

- elongation (Fig. 6.10a),
- irregularity (Fig. 6.10b), and
- composition (Fig. 6.10c).

Case I - elongation. This case is very common for nodular particles after plastic working; for example, rolling, pressing or axial tension. Elongation can be effectively measured using the following form factor (see Fig. 6.11):

$$f_1 = \frac{a}{b} \tag{6.5}$$

where: a and b are the length and breadth of the minimum bounding rectangle.

The shape factor f_1 can be also computed as the ratio of the maximum Feret diameter to the Feret diameter measured perpendicularly to it. The above defined shape factor gets the minimum value of 1 for an ideal circle or a square and greater values for other, elongated shapes. The greater the value of the shape factor, the more elongated particle.

Case II - irregularity. This case is also often met in practice. Many thermal processes, leading to equilibrium conditions, cause particles to become more and more perfectly spherical. Unfortunately, the shape factor f_1, defined by eqn. (6.5) is useless for irregularity assessment, as all the particles in Fig. 6.10b will have values very close to 1. A good solution to this case is given by one of the most popular shape factors:

$$f_2 = \frac{L^2}{4\pi A} \tag{6.6}$$

where: L is the perimeter and A is the surface area of the analyzed particle (see Fig. 6.11).

The shape factor f_2 is very sensitive to any irregularity of the shape of circular objects. It has a minimum value of 1 for a circle and greater values for all the other particles. This does not mean, however, that the shape factor f_2 is better than f_1 because f_2 is much less sensitive to elongation.

Case III - composition. This case can be treated as a kind of mixture of elongation and irregularity. This case is also met in real materials. An example can be the change in form of the graphite precipitates during transition from the nodular to flake cast iron. Any attempt to characterize this transition by one of the above defined shape factors will fail.

A solution very often used for such a case is computation of the weighted average of a few shape factors. Indeed, one can successfully manipulate the weights and get satisfactory results, but the result cannot be physically interpreted. If we have the elongation shape factor equal to 10, we know that the particle is quite long. If we have the irregularity shape factor equal to 3, we know that the particle is either extremely long or its boundary line is irregular and heavily curved. But we cannot interpret the following message: 30% of elongation + 50% of irregularity + 20% of number of holes gives the shape factor with the value of 3.47. There are hundreds of particles with completely different shapes that fulfill the above conditions. So, it is better if we can to construct an appropriate shape factor.

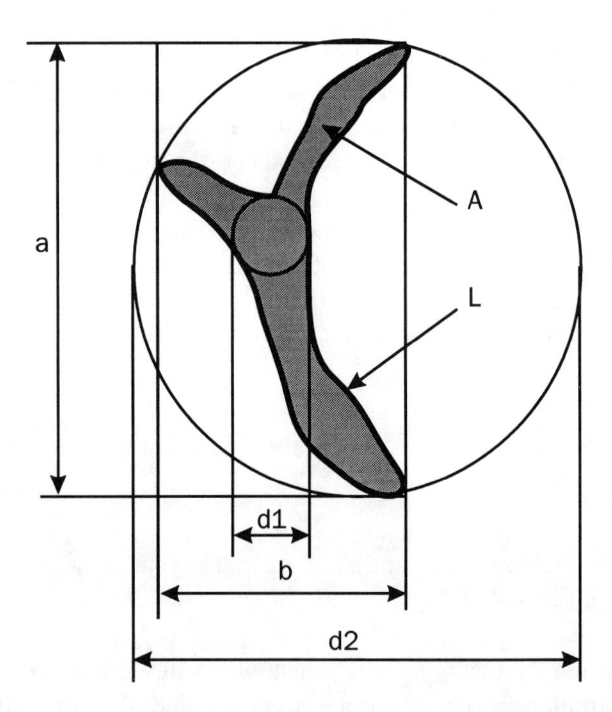

Fig. 6.11. Basic measurements used for the evaluation of the shape factors.

In the case of the shapes shown in Fig. 6.10c one can note the following rule: all these particles have approximately the same surface area, but as we move across the sequence to the right a smaller object can be drawn inside the particle. This observation leads to a definition of the following new shape factor (see Fig. 6.11):

$$f_3 = \frac{d_2}{d_1} \tag{6.7}$$

where: d_1 and d_2 are the diameters of the maximum inscribed and circumscribed circles, respectively.

The newly defined shape factor works really well. This shape factor can be applied to either elongated or irregular particles. However, as was mentioned earlier in this section shape cannot be precisely described by a single number. Therefore, the shape factor f_3 can be used for quantification of elongated or irregular particles, but cannot be used for recognition or *classification* of the particles. In other words, using this factor does not allow us to decide if the particle is elongated or irregular.

For classification purposes we have to use more than one shape factor. An example of a classification table is presented below. Note that this is only a very rough idea of classification, so cannot be directly applied in any real analysis.

Table 6.1. Shape classification table.

elongation	irregularity	analyzed shape
low	low	circular
low	high	compact, irregular
high	low	long and smooth
high	high	long and irregular

Very interesting results can be obtained if we apply the rules of fuzzy logic. In binary logic a particle can be regular or irregular. In fuzzy logic the same particle can be regular, irregular, almost regular, weakly regular, etc. Fuzzy logic allows us to measure how near a particle is to regular.

This concept is well illustrated in Fig. 6.12. We assume that the particles with elongation between 0.4 and 0.7 are thin. Particles with elongation less than 0.2 or greater than 0.9 are not thin. The rest of particles are partially thin and the percentage of thinness is illustrated by the plot in Fig. 6.12. Application of fuzzy logic has three main advantages:

- fuzzy logic enables quantification of particles which do not fit the assumed templates. In binary logic these particles are outside classification
- in fuzzy logic we can easily apply the above-mentioned method of using a weighted average for evaluation of the shape factors. Even more sophisticated rules can also be applied. The advantage of fuzzy logic lies in the fact that the results are always in the 0-100% range, so there is no longer any difficulty in physical interpretation of the results
- the results offered by fuzzy logic are relatively close to our human way of quantification.

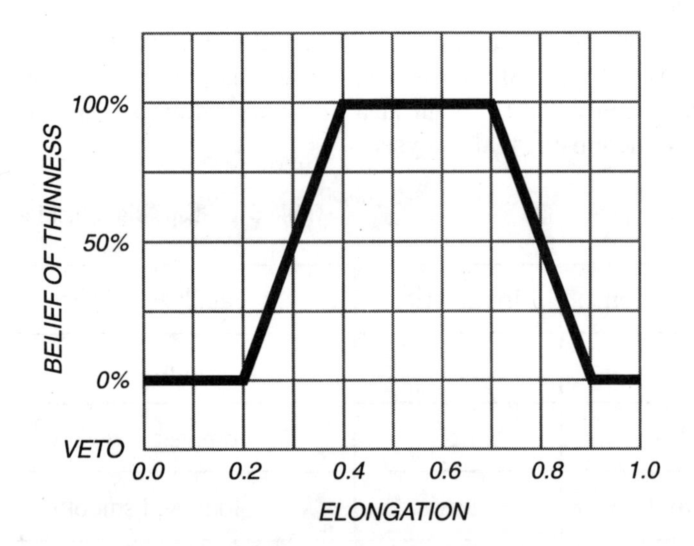

Fig. 6.12. Schematic illustration of the idea of fuzzy logic.

An example of the application of the fuzzy logic to shape analysis is shown in Fig. 6.13. This figure contains a collection of real graphite shapes, detected from cast iron of various grades. Every particle is described by two numbers: the value of f_3 (eqn. 6.7) and the circularity rating, expressed by percentage. This last value was computed using fuzzy logic. It is clearly visible that four particles at the left side of Fig. 6.13 are recognized as fully circular. The particles in the middle of the picture are partially circular whereas graphite flakes (right side of Fig. 6.13) are judged as absolutely not circular.

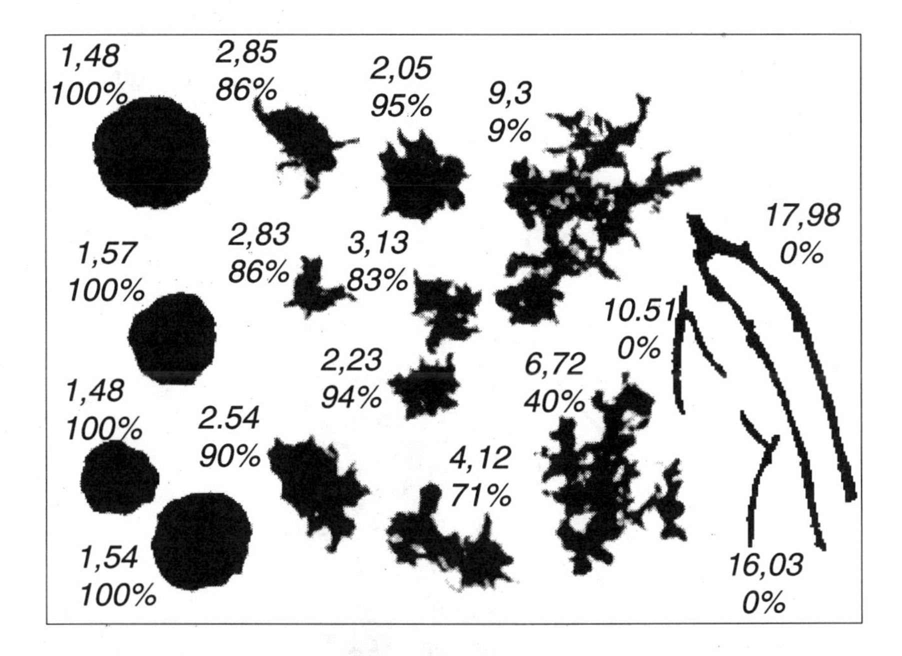

Fig. 6.13. Application of fuzzy logic to classification of the graphite particles. Upper numbers denote the value of shape factor and lower numbers (%) inform how well the particle fits as a circular shape.

6.5 Grain size

Grain size is probably the most important parameter for single-phase ceramic or metallic materials. This is also one of the basic measures for quantitative characterization of multiphase materials. A lot of research work[15,17,18, 21,24,53,54,57,65,91,100] has been done in order to evaluate the grain size and has proven that this problem is very complex and no universal solution can be found. Below you will find a brief overview of grain size measurement methods as well as some comments on computer-aided measurements devoted to grain size evaluation.

Grains (Fig. 6.5) are three-dimensional objects and theoretically their size can be effectively described by one of the following parameters:

- number of grains per unit volume N_V,
- mean volume of a single grain $\bar{v}$ or its distribution,
- surface area of grain boundaries per unit volume S_V, or
- any dimensions, but really measured in three dimensions.

In the case of real materials we have no possibility of evaluating in a reasonable time or for an acceptable price the true distribution of grain volume. On the basis of measurements performed on flat sections we can estimate the N_V and S_V values. The appropriate equations will be shown below. The number of grains per unit volume and the mean volume of a single grain are bound together by the following simple equation:

$$\bar{v} = \frac{1}{N_V} \tag{6.8}$$

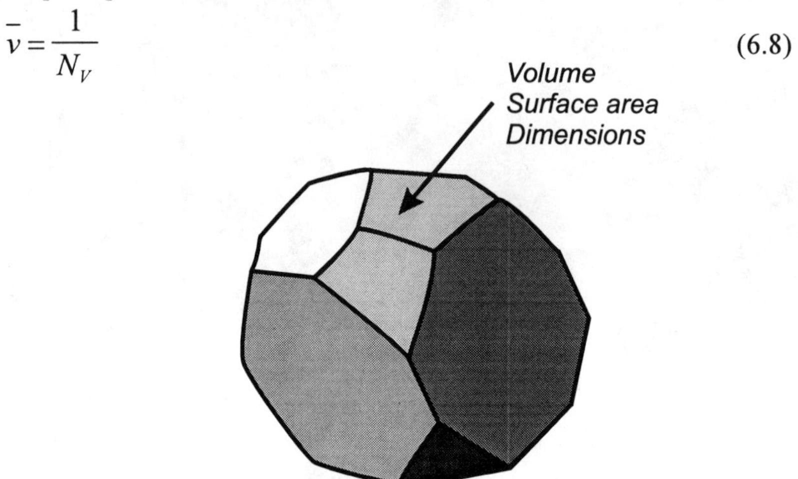

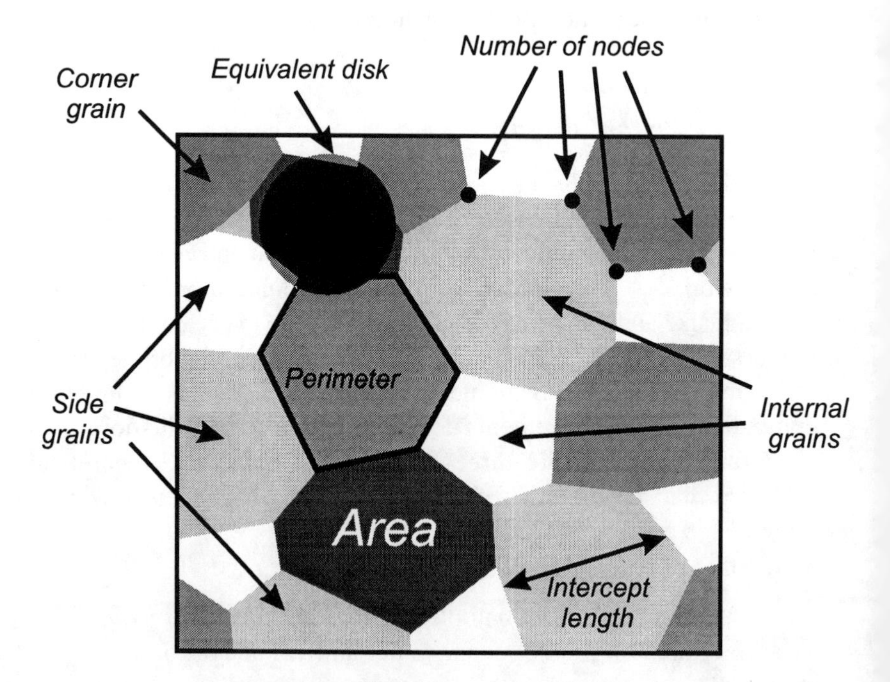

Fig. 6.14. Grains and their basic measures. We need to characterize the three-dimensional grains (top) on the basis of analysis of flat sections (bottom).

In the case of flat sections (usually this is just a material supplied for image analysis) we can measure many parameters related to grain size, for example (see Fig. 6.14):

- number of grains per unit section area N_A
- distribution of the section area $\bar{a}$ (the mean area of a section is just a reciprocal of the N_A value)
- equivalent disk diameter - this is the diameter of a disk whose surface area is equal to the surface area of a grain section. The mean equivalent disk diameter is one of the oldest measures of grain size. Currently this parameter is criticized because a circle is not a good model of grains, which are observed in the section plane as polygons filling the whole surface. We cannot cover the whole surface by circles
- number of triple points (nodes) per unit area P_A
- intercept lengths l.

The number of grains per unit area N_A can be evaluated using the Jeffriess method, known from classical stereology:[100]

$$N_A = N_i + 0.5N_b + 1 \tag{6.9}$$

where: N_i is the number of internal grains and N_b is the number of side grains. Here it is assumed that four corner grains together constitute a single grain.

It should be stressed that in the case of computerized methods a much better solution is to use one of the techniques for counting objects, described in Section 6.3 (see Fig. 6.6). Another very clever method is based on counting the number of triple points[82] (see Fig. 6.14):

$$N_A = 0.5P_A \tag{6.10}$$

where: P_A is the number of triple points (nodes) per unit area.

Triple points, assuming the width of the grain boundary lines is only a single pixel, can be easily detected using the HMT transformation and all the structuring elements containing three "1" and five "X".[84,114] Also, without any special difficulty, we can measure the mean intercept length. This can be done by counting the line-entry points for orientation angle 0° (see Section 6.2). Next, we should divide this number by the number of lines in the image (vertical dimension of the image expressed in pixels) and multiply the result by the horizontal dimension of the image. Now, we can easily compute the S_V parameter as

$$S_V = \frac{2}{\bar{l}} \tag{6.11}$$

where: $\bar{l}$ is the mean intercept length.

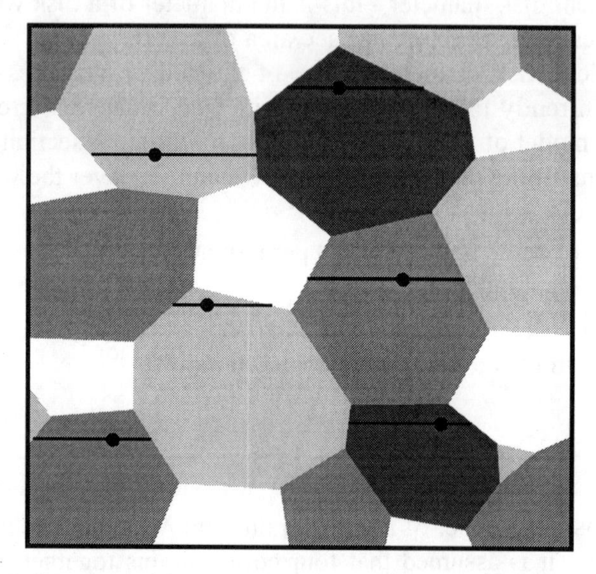

Fig. 6.15. Point-sampled intercept length method, schematically. See text for detailed description.

Estimation of the number of particles per unit volume N_V on the basis of flat sections is almost impossible without some assumptions concerning the shape of objects analyzed. It is possible, however, to adapt the point-sampled intercept length method, originating from pathological research.[37] In this method we use a set of statistically randomly distributed points, marking the position of intercept lines which, in turn, can be randomly oriented. In the case of random structures, the intercepts can be also parallel (see Fig. 6.15). For convex particles we get a very simple, however unbiased and statistically exact, equation for computation of the mean grain volume:

$$\overline{V} = \frac{\pi}{3}\overline{l^3} \tag{6.12}$$

where: l is the length of the point-sampled intercept.

So, we have a few measures of grain size. Among them the most important and frequently used are (symbols explained earlier in this section: $\overline{V}$, N_V, $\overline{a}$, N_A, S_V). Unfortunately, they cannot be used interchangeably, as there is no exact relation between all of them. Therefore, in order to choose an appropriate measure, one should take into account a number of different factors:

- if there are any archival results of measurements, the new measurements should be comparable with the old ones. The value of a large amount of archival material in many cases cannot be overestimated and it is often better to get results that are theoretically less correct but possible for comparison with a large existing database
- the measures chosen for our research should be feasibly precise and computed quickly by our analysis system. Some measures may be unavailable and some others can be done almost immediately without any difficulty
- mean values are easy for further analysis but can be sensitive to subtle changes under consideration. Therefore, it is a good practice to use for advanced analysis parameters that can be measured in terms of their distribution. A typical example is the measurement of particle section areas. These measurements are very fast and simple - a computer measures the surface area by a simple count of the pixels belonging to a given object, so they can be easily done, offering a large number of data suitable for advanced statistical analysis
- in general, all the stereological equations require IUR (independent, uniform and random) distribution of the structural constituents and a large number of observations, assuring unbiased estimation. These factors have to be taken into account prior to any measurements.[28,37,82,100,104]

A very interesting problem is connected to the meyhod of interpreting distribution data. Let us assume that we need to analyze the size distribution of grains, as shown in Fig. 6.16. The most simple and natural seems to be application of the grain area as a measure and number-weighted distribution (gray bars in the plot in Fig. 6.16). Now, we should explain the term *number-weighted*. Number-weighted means that the particles are counted proportionally to their amount, measured by numbers. In other words, this distribution illustrates *how many* particles of a given size we have. However, another concept is possible and, moreover, this second concept is at least equally valuable.

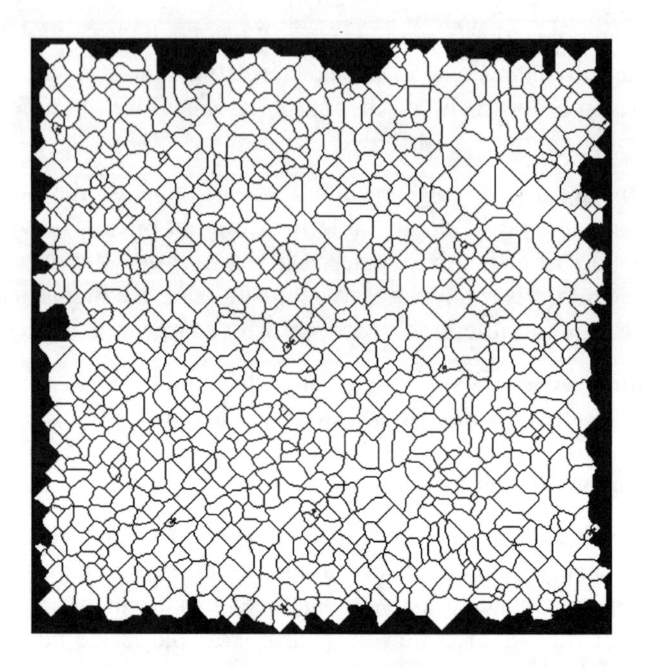

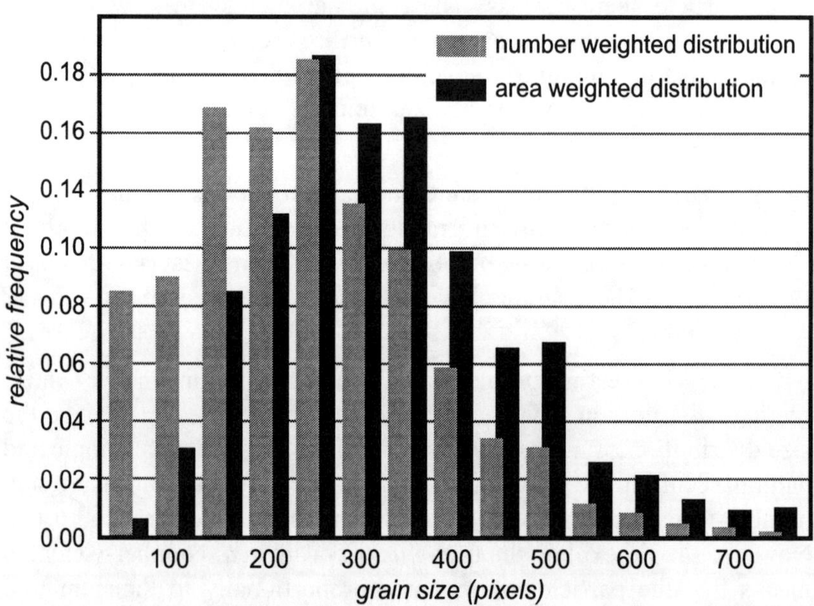

Fig. 6.16. A simulated single-phase microstructure (top) and two different size distributions (bottom): number-weighted (gray bars) and area-weighted (black bars). The latter seems to be more informative.

If we analyze the probability of hitting a particle with a single, randomly thrown point we will conclude that this probability is proportional to the area of this particle. Similarly, many processes in the material are related to the particle volume rather than to the number of particles. Thus, a distribution in which the weighting parameter is area instead of a number seems to be more adequate to relate to the physical properties of the material. Note (Fig. 6.16) that the area-weighted distribution is shifted in comparison with the number-weighted distribution towards larger particles. This phenomenon is even more pronounced in the case of a bimodal grain size distribution. The two maxima are usually hardly visible in the number-weighted distribution but very clear in the area-weighted distribution.[91]

6.6 *Gray-scale measurements*

All the measurements discussed in the previous sections refer to binary images. This allows one to count the number of features, measure size and shape characteristics or collect data for analysis of spatial arrangement of the structural constituents. On the other hand, binary representation loses all the gray-scale information, characterizing numerous important properties of the material, like, for example, microsegregation, inhomogeneity of phases, texture, etc. This can be done using gray-scale measurements.

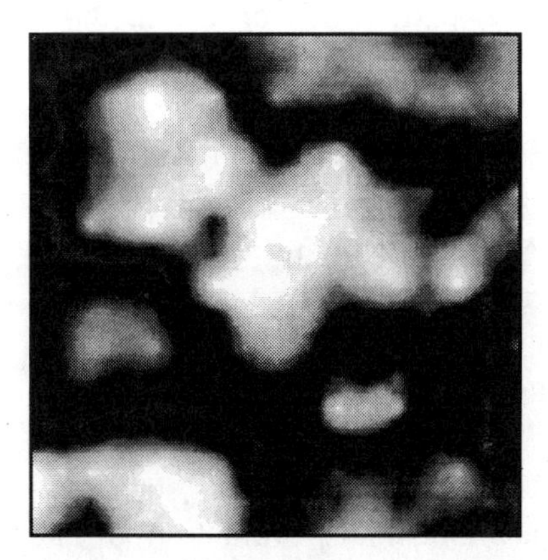
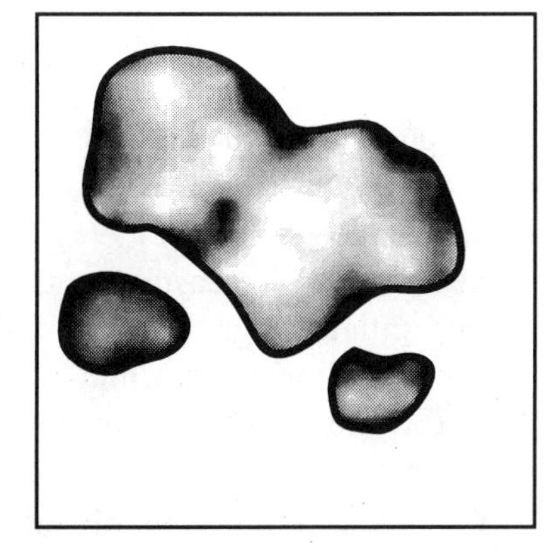

Fig. 6.17. Gray-scale measurements, schematically. The gray image (left) is masked by a binary image (right). This operation produces separated gray-scale regions which can be effectively analyzed.

The main difficulty in any attempts to perform gray-scale measurements is that a computer does not separate objects in a gray-scale image as humans do. Therefore, for gray-scale measurements, the binary image is used as a mask (Fig. 6.17). Masking can extract separate gray-scale regions, which in turn can be numerically analyzed. In this way we can get the minimum, mean and maximum gray level as well as its variance for any particle. Appropriate filtering of the gray-scale image can give additional information on the substructure of these particles, not available in another way. This kind of measurement can be easily done in object-oriented systems.

Application of gray-scale measurements offers some unique properties, leading to interesting applications:

- maximum of the distance function built on the binary image, multiplied by two, gives maximum widths of particles
- analysis of the gray-level distribution offered by gray-scale measurements allows separation of subtly differentiated particles (for example, twins or sub-grains) which cannot be effectively separated using other methods.

6.7 *Other measurements*

Obviously, the contents of this chapter does not cover all the questions related to digital measurements in image analysis. In particular, we have not discussed inhomogeneity, orientation or periodicity. In-depth analysis of these features requires thorough stereological knowledge which lies outside the framework of this text. However, the methods of stereology can be relatively easily adapted to the needs of image analysis even if elaborated in the pre-computer era. Below you will find some guidelines which should help in preparation of digital measurements:

- always keep in mind the digital nature of computer images. If you need to make measurements in various directions you can rotate the initial image instead of rotating the measurement direction. This solution should minimize the measurement errors. You can also try to convert the image into vector graphics if your software can perform measurements of vector graphics objects
- the weakest point in digital measurements is evaluation of the length of curved lines. Avoid such measurements if possible
- some advanced stereological methods (for example, the method of vertical sections)[2,54,115] use curved test lines. These methods can be effectively adapted to the needs of digital measurements. The sim-

plest solution is to perform measurements in many different directions and compute the final result as the weighted average of results for different orientation angles. In the case of a cycloid applied in the method of vertical sections, the sine function is used as a weight.[2,107]

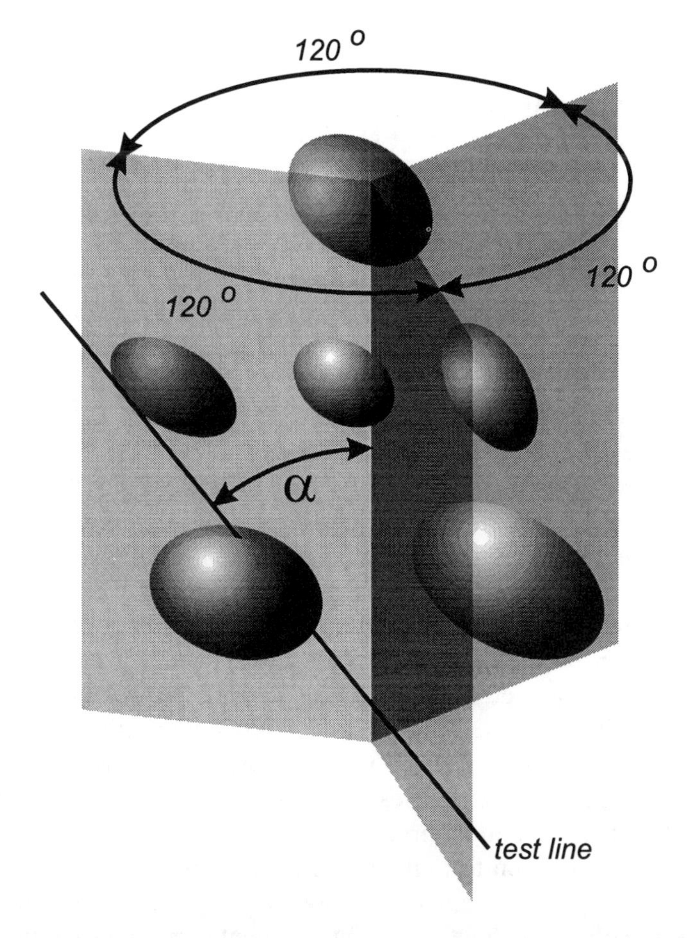

Fig. 6.18. Three parallel sections can give satisfactory results if the material exhibits banding or orientation. See text below for details.

- in general, stereology assumes that the microstructural features are IUR, this means *independently, uniformly randomly* distributed. This assumption is not fulfilled (or fulfilled only partially) by the majority of real materials. This problem can be easily solved by using IUR section planes and test lines. Unfortunately, this solution is simple only in theory. However, application of three section planes, rotated by 120°, is a sufficient remedy in most cases (see

Fig. 6.18). In the case of some stereological parameters evaluated on the basis of test lines, this solution requires special analysis of the results or nonlinear test lines, mentioned above. In order to simulate IUR test lines, you should apply many test directions and multiply the results obtained for any test line by $sin(\alpha)$. A more detailed description of this method[54,115] can explain its theoretical principles but the above short description is sufficient for practical applications

- many stereological methods use as an input parameter the number of intersection points per unit chord length. You can build an intermediate image containing the test lines and find the points of intersection with the analyzed particles using a logical AND transformation. A more elegant solution is to use the HMT with an appropriate structuring element (see Section 6.2, Fig. 6.4), as the sum of projected lengths per unit area L_A is equal to the number of intersection points unit chord length P_L.[82] The only condition is that the intersection points should be counted in a direction perpendicular to the projected lengths
- coordinates of some unique points, for example, centers of gravity, can be effectively applied in analysis of the spatial distribution of particles, inhomogeneity, clustering or periodicity
- if you have to quantify any more sophisticated property of the material microstructure it is a good practice to check if this feature can be emphasized with the help of some transformations. A good example can be the autocorrelation function[80] which can characterize orientation (see Fig. 6.19). We can binarize the autocorrelation image in such a way that binarization will give an object in the center of the image. Its shape will give the overall orientation of the initial image. For example, it will be elongated in the direction parallel to the orientation direction.

The final conclusion from the above analysis is that we can easily apply almost all of the methods elaborated within the framework of classical stereology, without taking into account specific properties of digital measurements. Some of these methods require modifications in order to avoid possible numerical errors or speed-up the process. Simultaneously, image analysis offers new opportunities not available for classical stereology. These new possibilities are visible, first of all, in the image processing stage. However, some of them, for example, gray-scale measurements, can be noted also at the analysis and interpretation stages. Last but not least, the ease of performing measurements, leading to a great number of data allowing distribution analysis, cannot be overestimated.

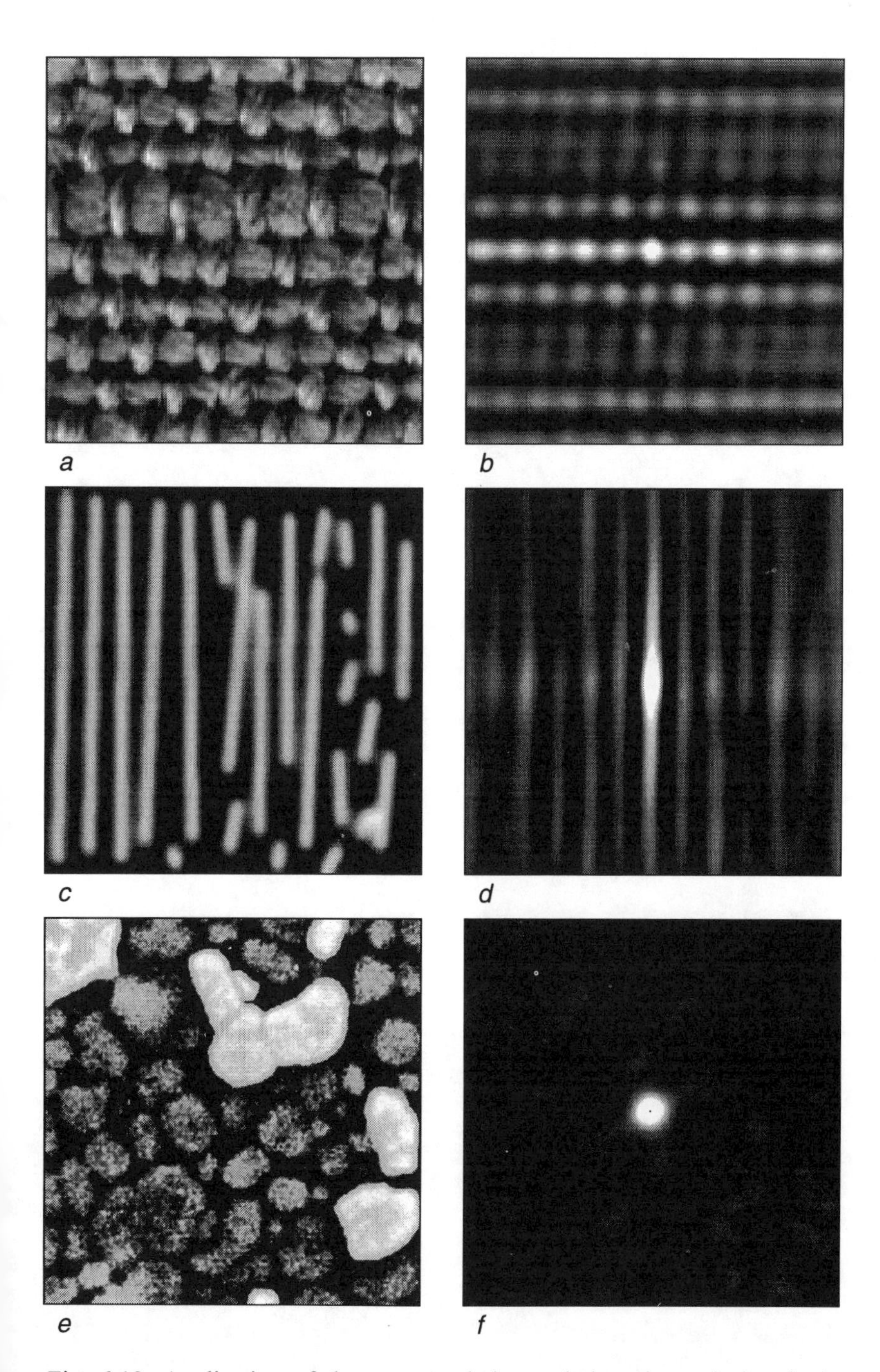

Fig. 6.19. Application of the autocorrelation technique in analysis of the overall orientation of the image. Bright areas in the autocorrelation images (right column) can be extracted using binarization and their shape informs about orientation.

Chapter seven

Applications and case histories

In classical textbooks on image analysis we can find description of the nature of various procedures or algorithms and their properties. This solution is attractive for both specialists and newcomers interested in thorough study of the discipline. However, this is unacceptable for researchers looking for a quick, possibly simple and effective solution to their individual problems. In order to meet their needs this book is organized in another way:

- we go through real problems and show how to solve them rather than discuss the properties of algorithms and operations applied
- we try to explain why the given sequence of operation is successful and anticipate how this solution could be adapted to other, more or less similar problems
- all the problems discussed originated from materials science. There are no examples from biology and medicine, remote sensing, criminology or military applications.

From the above characteristics it is clear that the whole text is entirely devoted to applications in materials science. However, even in this user-oriented style there are some drawbacks. For example, the concept presented above is suitable for a person who already has a problem to solve. Furthermore, this is absolutely insufficient for anyone wanting to learn what potential problems can be solved using image analysis. Therefore this last, short chapter is devoted to applications. We can list three groups of main application directions in materials science, which will be discussed in the next sections:

- quality and routine control
- simulation
- research and case histories.

7.1 Quality and routine control

There are some characteristics of the microstructure that control most of the material properties. Among them we can list volume fraction, grain size or orientation. Characterization of these quantities may be required in exposed applications, for example, turbines, pipelines or automotive industry and appropriate tests should be done during routine quality control. The needs of this structure-oriented quality control are simultaneously simple and difficult:

- usually only relatively simple parameters, like inclusion contents or grain size should be measured. Appropriate measures are well defined within the framework of stereology and almost any image analysis software is capable of returning appropriate results. So, this looks to be simple
- the analysis should be quick, possibly automated, ensuring high repeatability of the results and minimized bias introduced by the operator. Many inter-laboratory tests have proven that completion of all of the above criteria may be extremely difficult.

In order to achieve acceptable results, all the procedures used have to be thoroughly tested and described in detail. Thanks to the intensive work of standardization committees this work has been already done for some parameters and resulted in appropriate standard procedures.[89-91] The existence of standard procedures does not mean that their practical application is always smooth and painless. However, the main problems are connected with proper sampling and specimen preparation rather than with the final image analysis.

As you have probably noted, image analysis tools are both very flexible and sensitive to fine changes in their tuning. This flexibility enables, for example, detection of grains in images of various type and quality (see Section 4.1) but simultaneously requires intuitive selection of the transformations. Standard procedures should avoid the use of intuition and appropriate sequences of transformations should be predefined. Additionally, small variation in the threshold level (see Section 2.4) can result in dramatic changes in the measured contents of phases to be detected, especially in the case of any scratches in the specimen image. This property can also considerably complicate preparation of any standard practice.

In order to neutralize or at least minimize the above-mentioned factors, standard procedures apply the following techniques in the case of automatic inclusion assessment of steel:[90]

- special attention is paid to specimen preparation. Specimens should be cut with care in order to avoid unnecessary deformation. Next, they should be polished using appropriate silicon carbide or aluminum oxide papers under a plentiful stream of water. Ultrasonic cleaning is recommended after grinding. For final polishing, diamond pastes are recommended as well as an absolute minimum polishing action

- the polished specimen should conform to very high quality standards. The specimen quality should be tested using microscopic observation at magnifications 100x and 500x, using differential interference contrast
- extremely high standards of specimen preparation enable further automatic analysis. The contrast between inclusions and the polished matrix is high enough for automatic thresholding. Good specimen preparation produces very thin boundary lines separating the inclusions from the matrix. Therefore, a small alteration in the threshold level does not introduce visible differences in the amount of the detected phase. This is noteworthy, however, that even in this extremely well-defined case a semi-automatic method, based on alternating between the live video image and the thresholded image, is recommended for optimization of the threshold level.[89]

In the case of grain size analysis, the situation is even more difficult. The specimen has to be not only perfectly polished but also correctly etched. The number of etching mixtures is still growing, so one should be very careful when deciding which reagent should be chosen. There are specialized reagents suitable for a very narrow group of materials, having the appropriate chemical composition and technological history. In general, such etching techniques are preferred that ensure further detection with minimum possible difficulty.

In other words, the standardized procedures say: do not use extensive image processing. Rather, try to prepare your specimens in such a way that further detection can be done without any problems, even by a simplified software. Such specimen preparation requires enormous experience and extensive knowledge of laboratory techniques, as well as the properties of the materials being tested. Clearly, such knowledge is not the subject of this book. We do not analyze specimen preparation techniques but focus on cases in which appropriate specimen preparation is difficult or even impossible. For example, in the case of many ceramic materials we have no technical aids for clear detection of the existing grain boundaries. Successful examination of such difficult cases requires another type of knowledge, probably more intuitive and often unpredictable and the aim of this book is help you to acquire this skill.

Obviously, there are numerous cases when we should elaborate the rules of routine quality control in spite of the lack of appropriate standardized procedures. In such a case we should apply the guidelines presented in the next sections.

7.2 *Simulation*

The next field of extensive use of image analysis is simulation. In recent years simulation has been used very frequently: we can simulate car accidents, financial operations, weather changes, etc. Using image analysis tools we can simulate various microstructures and some processes leading to microstructure alteration. There are some general problems which can be solved using image analysis-based simulation:

- new stereological tools are often based on statistical analysis or very complex theoretical analysis. Even if the procedure is free from systematic bias, it is especially difficult to fix theoretically its precision and necessary number of counts until convergence. By contrast, this can be easily done using the simulated structures. Their advantage lies in the fact that any simulated structure is well defined and we can exactly evaluate all the necessary parameters. The exact values can be used for comparison with the results returned by procedures being verified, thus giving us the necessary information concerning their precision and bias. If we use any real structures for the same purpose, we never know what the real value of the estimated parameters is
- simulation of materials structures is another field for extensive application of image analysis. This is very effective in the case of modeling grainy structures. Such simulation can be easily performed using some predefined markers, playing the role of grain seeds (these can be, for example, randomly generated points). The final structure can be easily built using the SKIZ transformation. In order to create simulation closer to reality we can interrupt the process at any stage and add some new grain seeds. By the skillful addition of new grain seeds, this modified procedure allows us to simulate grains to look very close to reality. We can simulate the balance between the rate of new grain seed formation and crystal growth (see Fig. 2.21). Comparison between the simulated and real structures helps in evaluation of the structure formation processes (in this particular case, this is solidification). More advanced models can lead to creation of new technological processes, resulting, for example, in intentionally oriented or more homogeneous structures
- simulation of various processes is the third major application discussed in this section. Simple removal of the smallest grains with subsequent SKIZ can roughly approximate the grain growth process. Other, more complex models, can be effectively used for in-

vestigation of diffusion, plastic deformation, corrosion or fracture processes. An illustration of such a simulation will be presented below.

In the example below we will analyze simulation of the fracture path in a nodular cast iron. It is assumed that the fracture growth follows a simple criterion of the minimum energy necessary to create a fracture path. Due to very small cohesion forces between the metallic matrix and graphite nodules we can treat them as pores. So, the energy necessary to cut graphite nodules is assumed to be equal to null. This leads to the following geometrical rule: the growing fracture should minimize the length of the overall path through the matrix and the length of the fracture path through graphite is insignificant. Now, we will see how this simple rule can work in the image taken from a real cast iron structure:

- the crack will be simulated within the microstructure shown in Fig. 7.1a. The starting point is situated at the top edge and indicated by an arrow. The crack should propagate towards the bottom edge of Fig. 7.1a
- the algorithm should ensure proper crack propagation direction. So, in order to prevent erroneous crack growth we define appropriate mask. This mask has a triangular shape with one vertex placed at the starting point of the crack. The set of graphite nodules reduced to the area of the mask is shown in Fig. 7.1b
- now we go to the main body of the whole process - simulation of the growing crack front. This is done by summation of dilations. The first one is dilation of the starting point indicated by an arrow in Fig. 7.1a. This forms the intermediate image to which every next dilation is added. We can call this intermediate image *distance image* because the gray level of any point within this image is proportional to the distance from the crack front. Obviously, the starting point of this simulation process is the most distant from the simulated crack front (Fig. 7.1c)
- the postulated null energy necessary to go through the graphite nodule is converted into a simple geometrical rule: if only the simulated crack front touches any nodule, this nodule is added to the crack front. Fig. 7.1d illustrates how the existing nodules influence the shape of the crack front
- the above described process of crack front growth is repeated until the growing front reaches the edge of the image opposite to that containing the starting point

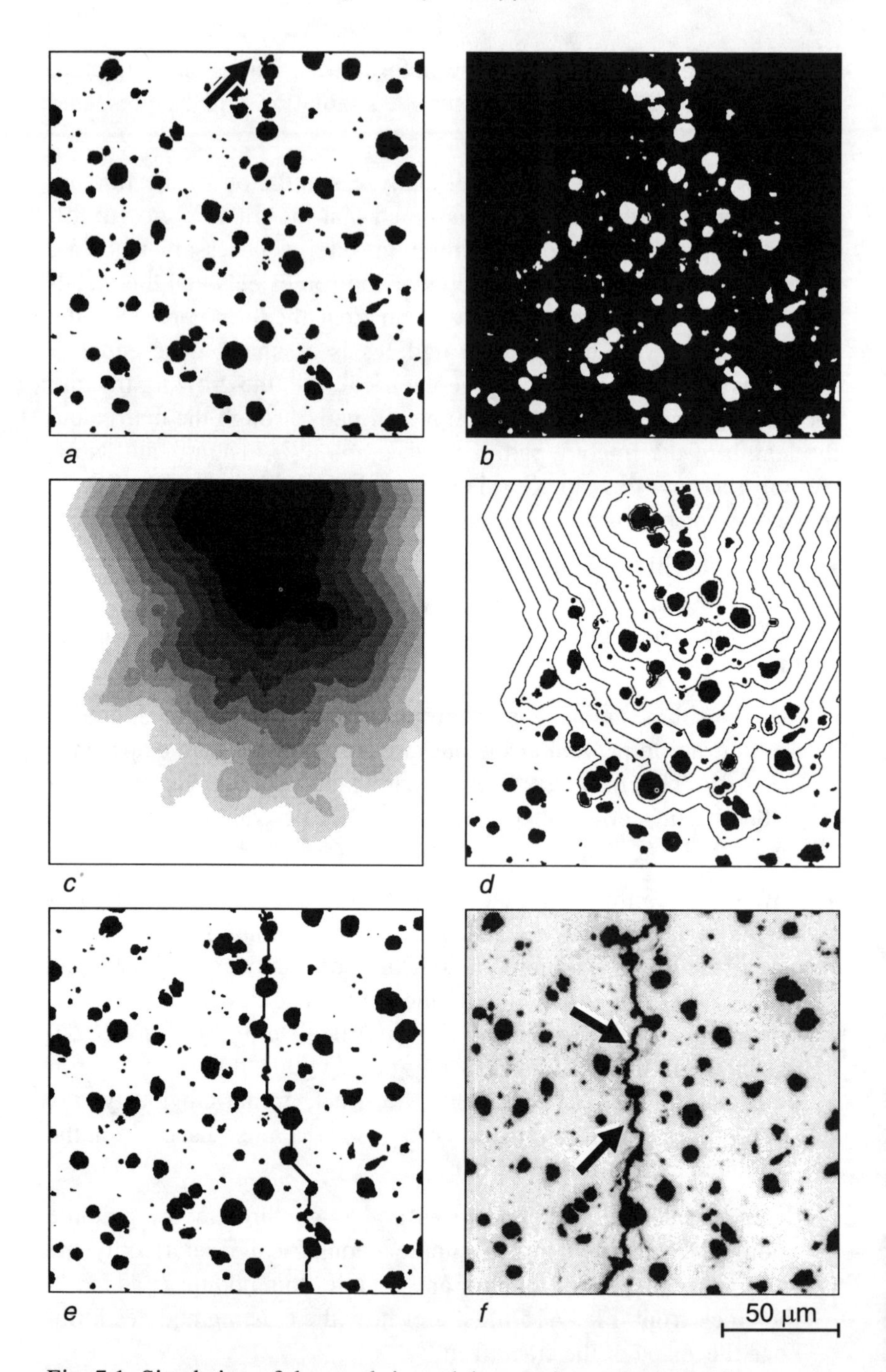

Fig. 7.1. Simulation of the crack in nodular cast iron. The initial image with the starting point indicated by an arrow (a), masked nodules (b), intermediate distance image (c), illustration of crack growth mechanism (d), the final simulated (e) and real (f) fracture paths.

- the final detection of the simulated crack path is done by watershed detection within the distance image. The resulting path should next be cleaned of unnecessary branches
- the final simulated crack can overlay the microstructure (Fig. 7.1e).

If we see the simulated crack with the real one (Fig. 7.1f), we see that there are significant differences between them. This does not mean that our simulation technique is totally wrong. This is rather an effect of two-dimensional nature of our image. In reality, the crack direction is affected by nodules located in 3-D space, so some of these nodules are not visible - we can observe only some unexpected bends, like those indicated by arrows in Fig. 7.1f. However, it is noteworthy that a very similar algorithm, applied to a more brittle material - porous graphite used in nuclear power plants - gave very satisfactory results.[48,102] Simulation techniques allow also objective testing of various methods of grain size inhomogeneity measurements.[17,18,57]

Image analysis and simulation are tightly bonded together. For example, it is impossible to decide if grain size restoration algorithms contain elements of simulation, or not. The remarks put in this section have mentioned some of the possible application directions of simulation techniques. We will not go further into this area. You should use your own imagination to formulate appropriate models of processes observed in materials as discussion of the rules of proper modeling is outside the scope of this book.

7.3 *Research and case histories*

Fortunately for research laboratories there are still numerous, practical problems to be solved. Among the variety of questions suitable for answering with the help of image analysis, we can find the following, given here as examples:

- we have two pieces of any material and need to judge if their microstructures are identical or not
- we have to develop a stable technological process leading to given properties of the final product and we get a large scatter of results. Therefore, we try to check what features in the microstructure are unstable
- we are looking for a subtle phenomenon difficult to detect (this can be, for example, the beginning of formation of the precipitation network) and need objective tools for microstructure examination
- we need to quantitatively characterize the microstructure evolution, for example, during long-term use of any power plant installation

- we need to explain the reason for any failure and, if possible, exclude material-dependent factors
- we are looking for any structure-property relationship and need to find the microstructure factors that controls the properties, etc.

So, it is clear from the above list that the number of possible applications is practically unlimited. There are so many materials, technological processes and their functions that this is impossible to deal with all of them or even with a representative subset. Instead, we can formulate some guidelines, which should be helpful in solving new problems:

- try to judge if the problem is suitable for image analysis. It happens quite frequently that we are asked for image analysis without prior preliminary studies indicating that the problem should be examined using image analysis tools
- check if you can give a manual way so these features can be automatically detected. In the other case, you will most probably fail when constructing the algorithm. It is possible that you will accidentally find the proper solution; however, such a fortunate case happens very rarely
- decide what structure parameters can be helpful for solving the particular problem. Keep in mind that simpler parameters usually require less image processing and, consequently, the error of subsequent measurements should be minimized in this way
- ensure that your images are of the best possible quality. Many of our colleagues think that computers are capable of processing images of worse quality than necessary for manual analysis and bring really poor images. Avoid such cases and always check if another visualization technique can offer better images
- if the problem is absolutely fresh to you, try to get as much practical knowledge concerning this case as possible. Maybe you will note that there are some similarities with other problems, already successfully solved by you
- if you have no idea what to do, check the demonstration algorithms in your software. The demos are usually well optimized and you can find many excellent ideas in this source of knowledge
- when you see that your work is going "nowhere" stop, and start from the very beginning. Building anything from scratch is always faster than modification of a poor idea
- if none of the above remarks works, skip to other work and do not trouble with it any longer. The break should be at least one night long. This advice really helps solve extremely difficult problems!

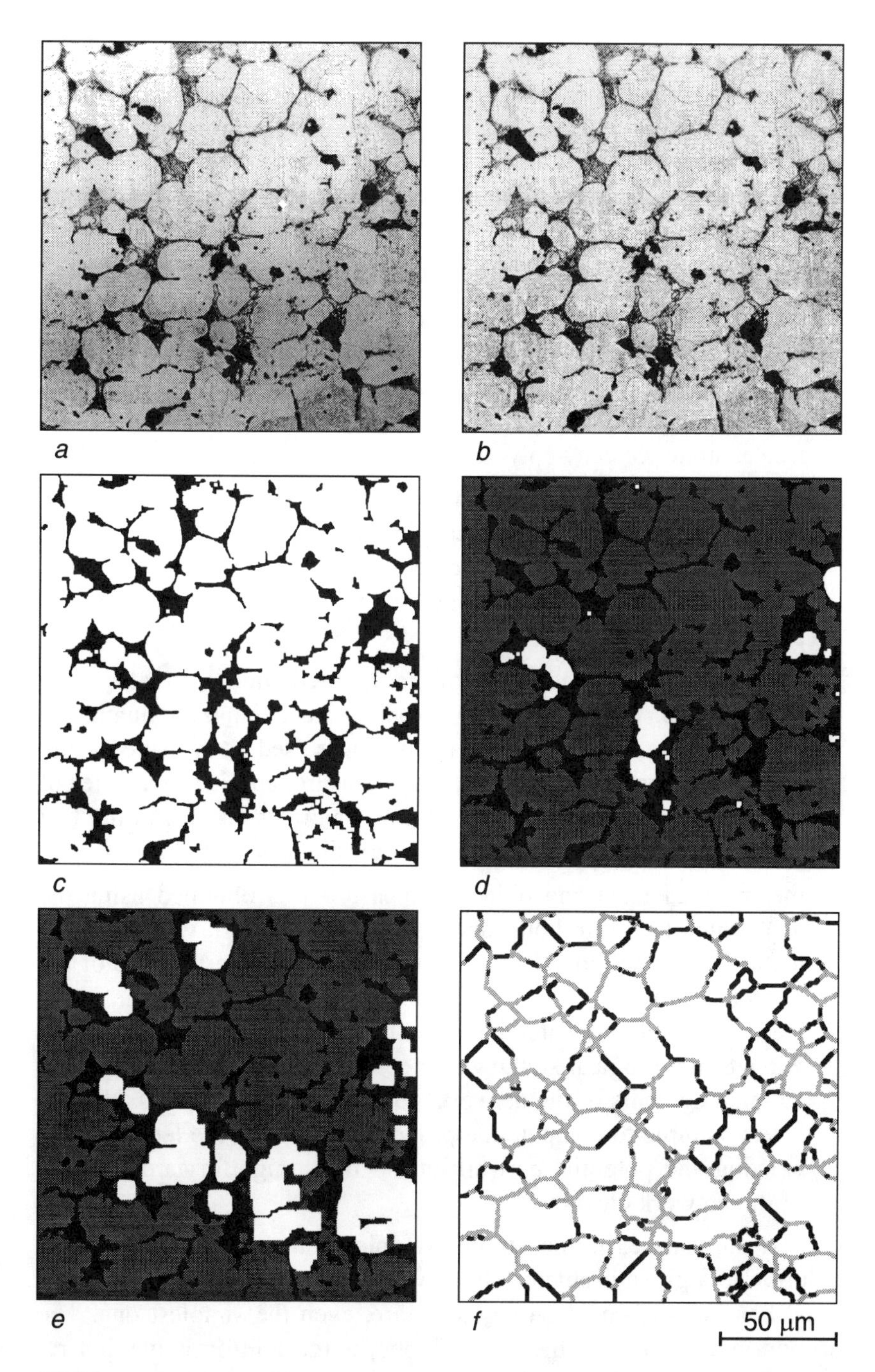

Fig. 7.2. Analysis of the eutectic network in the sintered stainless steel: initial image (a), image after shade correction (b), detected pores and eutectic (c), closed loops detected by small (d) and larger (e) closings and the potential eutectic network (gray) with missing parts indicated as black lines (f).

An example of a case history is presented in Fig. 7.2. During investigation of agglomeration of a sintered stainless steel, some addition of boron was used to minimize the porosity by formation of the eutectic phase. The goal of the investigation was to determine the ratio of eutectic network formation. The following analysis was performed:

- the first step was removal of the shade from the initial (Fig. 7.2a) image. The corrected microstructure is shown in Fig. 7.2b
- there are many pores in the material (black spots in Figs. 7.2a and 7.2b). For our analysis, however, the pores can be treated as a part of the eutectic network. Thus, we can get easily the image of the eutectic (with pores) by simple binarization. In order to get hole-free eutectic we can apply small closing (Fig. 7.2c)
- the progress in formation of the eutectic can be now easily measured by measuring the area occupied by this phase. Unfortunately, it turned out that area measurements were not sensitive enough to the progress in eutectic network formation. Therefore, another approach was introduced
- Closing applied to the image of the detected eutectic creates some closed loops. These loops can be easily detected and the number of loops or the area occupied by them can be used as measure of network development (Fig. 7.2d). This measure, applied to a series of various sinters, proved its usefulness. Its drawback lies in the fact that the results depend on the size of closing (see Fig. 7.2e)
- the most advance and objective measure was obtained using the algorithms for grain boundary restoration, based on watershed detection (see Section 4.1). Segmentation of the binary image of the microstructure (Fig. 7.2c) and subsequent SKIZ produce an idealized image of the potential eutectic network (gray lines in Fig. 7.2f). Simple logical AND with the initial binary image reveals the lacking elements of the networks (black lines in Fig. 7.2f). Now, it is very simple to compute the proportion between the length of the existing and potential network lines - the straightforward measure of the network closure.

To summarize, in this simple example we have three levels of the solution to a given problem. The lowest level (measure of the eutectic area) can be performed on any software, even the simplest one. The second level (counting the closed loops) is more informative, but requires much more knowledge and experience. The third level (based on segmentation and the SKIZ transformation) is clearly the most difficult one and simultaneously offers precise, quantitative description of the process being investigated. It should be pointed out that all of these solutions can be acceptable, depending on the required accuracy of analysis.

Now we will analyze a series of exemplary applications of image analysis in materials science. These examples are taken from the literature and therefore reflect various levels of experience, laboratory equipment and a large variety of materials and phenomena under consideration. Thus, this short review should inspire the reader to experiment with new, original applications:

- *study of the abnormal grain growth in sintered alumina*:[60] the measurements were carried out on digitized images of grain boundaries. In order to avoid the extensive image processing necessary to restore the lost boundary grains, the manually prepared images were used as input data. The results indicated that the process of sintering is accompanied by significant abnormal grain growth. This phenomenon was related to the segregation of impurities
- *directional fiber analysis by mathematical morphology*:[64] quantification of the fiber orientation is very important in various applications. A new algorithm, based on advanced mathematical morphology operations, is presented. One of the key points in this work is application of the directional erosion by a pair of points. The new algorithm enables analysis of individual fibers as well as detection of areas of similar orientation of the fibers
- *verification of empirical structure-property relationships in sintered carbides*:[75] thorough testing of various WC cermets observed in SEM and subsequently analyzed using image analysis tools, allowed the formulation of guidelines concerning specimen preparation, observation mode, optimum magnification, etc. The analysis performed allows recommendation of image analysis as a good tool for automatic quality control of the WC cermets
- *investigation of the drill wear mechanism*:[78] observation of the cutting surface of drills after lifetime tests seems to be one of the best sources of information on the wear mechanism. Unfortunately, the changes observed are predominantly of a quantitative character, without significant changes in the wear mechanism. The methods of quantitative fractography and image analysis can serve as a basis for creation of an expert test system
- *quantitative evaluation of multiphase materials*:[94] during SEM observation of non-asbestos friction materials with a complex chemical and phase composition the main problem is a poor contrast between the phases to be detected. The use of both BSE and

SE images allows correct detection and preparation of binary images for subsequent image analysis

- *automatic detection of twins*:[95] twins are an important part of the microstructure in many materials, for example, Cu alloys or stainless steels, and should be characterized in a quantitative way. Their automatic detection, however, is very difficult. Starting from the analysis of the distribution of angles between the grain boundary and twin boundary, the authors have developed an efficient algorithm, capable of correct detection of up to 90% of the twin boundaries
- *analysis of a high chromium steel for advanced power stations*:[117] this study has been performed on steels with high chromium contents (X20, P91, P92) designed for advanced power station. The material was examined using TEM; and image analysis tools enabled evaluation of such parameters as sub-grain width, dislocation density or particle size distribution
- *analysis of insulation mineral fibers*:[99] measurement of the diameters and lengths of fibers is important to the glass and mineral wool industry. The fibers are oriented in various directions and cross themselves. Correct segmentation in such conditions is a complex problem. Using advanced tools of mathematical morphology and specialized algorithms, the authors obtained good correlation between the manual and automated measurements
- *image analysis of austenite and carbides coarsening in a Fe-Mo-C steel*:[11] image analysis, similar to the example shown in Fig. 7.2, allowed the collection of quantitative metallographic data. The results were subsequently analyzed in terms of stereology and enabled verification of some microstructure evolution mechanisms
- *morphological study during a ceramic process*:[23] in this study the process of barium titanate ceramic formation was investigated. Three main stages of the process are studied using different methods of image analysis. The tools of mathematical morphology are extensively applied. The analysis enabled creation of numerous structure-property relationships which can be applied at the industrial level
- *automatic grain size measurement in low carbon steel by image analysis*:[56] this work describes a successful application of a fully automated image analysis in routine quality control. This work does not introduce sophisticated algorithms but proves that image analysis can speed up quality control and significantly increase the repeatability of measurements

- *modeling of the hydromechanical behavior of a fracture*:[36] this work is devoted to the analysis of hydromechanical behavior of fractures that is fundamental for understanding the effects of human engineering in fractured rock masses. Image analysis tools are applied here for enhancement of the images and for analysis of the fracture surfaces.

7.4 Concluding remarks

The previous section definitely closes the contents of this book. However, the problem of application of image analysis in materials science remains open. The author hopes that the tools and methods presented above will encourage you to apply image analysis in your everyday work. Moreover, solutions presented here are universal enough to be easily adapted to various materials not discussed here as well as to the needs of biology, medicine and other sciences. It is enough to look at Fig. 7.3 in which you can observe a surface of bread. This image is far from materials science but looks very similar to ductile fracture surfaces of metallic materials.

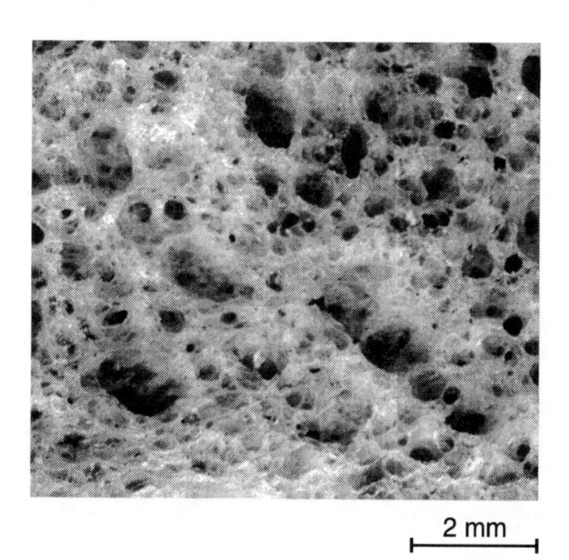

Fig. 7.3. The surface of bread is very similar to ductile fracture surfaces.

The structure shown in Fig. 7.3 can be surprising to people working with ceramics, metals or plastics. If you cooperate with specialists from other disciplines, this is a quite common situation and usually offers a lot of satisfaction. Obviously, new structures force development of new algorithms, which is laborious and time consuming. Moreover, you never knows if your skill and experience are sufficient to solve the problem.

When I started with the first draft of this text I also had a lot of fear about if I would finish it successfully. I did it. So, you should do the same.

References

1. Avérous, L., Quantin, J.C., Lafon, D. and Crespy, A., Morphological determinations of fiber composites, *Microscopy, Microanalysis, Microstructure*, 7, 433, 1996.
2. Baddeley, A.J., Gundersen, H.J.G. and Cruz-Orive, L.M., Estimation of surface area from vertical sections, *Journal of Microscopy*, 3, 259, 1986.
3. Barcelo, F. and Brachet, J.C., Image analysis measurements of prior austenitic grain size of 9Cr-1Mo martensitic steels in the function of the austenitization conditions, *Revue de Métallurgie*, 91, 255, 1994.
4. Beveridge, J.R., Griffith, J., Kohler, R.R., Hanson, A.R. and Riseman, E.M., Segmenting images using localized histograms and region merging, *International Journal of Computer Vision*, 2, 311, 1989.
5. Bjerregaard, L., Geels, K., Ottesen, B. and Ruckert, M., *Metalog guide. Your guide to the perfect materialographic structure*, Struers, Denmark, 1992.
6. Blanc, M. and Demagny, H., Caracterisation de l'anisotropie de structures granulaires par la mesure des roses des directions et des intercepts, *Journal de Microscopie et de Spectroscopie Electroniques*, 12, 109, 1987.
7. Bright, D.S. and Steel, E.B., Two-dimensional top-hat filter for extracting spots and spheres from digital images, *Journal of Microscopy*, 146, 191, 1987.
8. Broek, D., *Some contributions of electron fractography to the theory of fracture*, National Aerospace Laboratory NLR, The Netherlands, 1973.
9. Brolio, J., Draper, B., Beveridge, J.R. and Hanson, A., *The ISR: a database for symbolic processing in computer vision*, COINS Technical Report 89-111, University of Massachusetts at Amherst.
10. Bryant, J.D., Stereo-photogrammetry as applied to fractography. *Advances in Fracture Research*, Proc. ICF7, Houston, Texas, 3365, 1989.
11. Chaix, J.M., Yang, P.H., Ardin, M. and Durrand-Charre, M., Image analysis of austenite and carbides coarsening in a Fe-Mo-C steel, *Microscopy, Microanalysis, Microstructure*, 7, 387, 1996.
12. Chermant, J.L., *Les Ceramiques Thermomechaniques*, Presses du CRNS, Paris, 1989.
13. Chermant, J.L. and Coster M., Introduction à l'analyse d'images, *Journal de Microscopie et de Spectroscopie Electroniques*, 12, 1, 1987.
14. Chermant, J.L., Coster, M. and Gougeon, G., Shape analysis in R^2 space using mathematical morphology, *Journal of Microscopy*, 145, 143, 1987.

15. Chermant, J.L. and Coster, M., Granulometry and granulomorphy by image analysis, *Proc. 3rd Conference on Stereology in Materials Science STERMAT'90*, Szczyrk, Poland, 9, 1990.
16. Chermant, J.L. and Coster M., Granulometry and granulomorphy by image analysis, *Acta Stereologica*, 10, 7, 1991.
17. Chrapoński, J., *Analysis of applicability of stereological methods of evaluation of grain size in polycrystalline materials*, Ph.D. thesis, Silesian University of Technology, Katowice, 1997.
18. Chrapoński, J. and Maliński, M., Estimation of grain size, Part I-III, *Proc. Q-Mat'97, International Conference on the Quantitative Description of Materials Microstructure*, Warsaw, Poland, 215, 1997.
19. Chrapoński, J., Maliński, M. and Cwajna, J, The algorithm of three-dimensional reconstruction and its application in the materials science investigations, *Inżynieria materiałowa*, 17, 206, 1996.
20. Chrząszcz, R., *Application of image analysis in the inspection of the sintering process*, M.Sc. thesis, Cracow University of Technology, Cracow, 1994.
21. Coster, M. and Chermant, J.L., *Precis d'Analyse d'Images*, Presses du CRNS, Paris, 1989.
22. Coster, M., Chermant, J.L. and Prod'homme, M., Quantification of Ceramic Microstructures, *Proc. Q-Mat'97, International Conference on the Quantitative Description of Materials Microstructure*, Warsaw, Poland, 115, 1997.
23. Coster, M., Prod'homme, M., Chermant, L. and Chermant, J.L., Morphological study during a ceramic process, *Microscopy, Microanalysis, Microstructure*, 7, 407, 1996.
24. Cwajna, J., Quantitative structural criteria of steel quality control, *Acta Stereologica*, 13, 409, 1994.
25. Cwajna, J., Richter, J. and Szala, J., Quantitative metallography and fractography of modern tool alloys, *Proc. Q-Mat'97, International Conference on the Quantitative Description of Materials Microstructure*, Warsaw, Poland, 125, 1997.
26. Czarski, A., Satora, K. and Ryś, J., Quantitative description of the Armco steel structure, *Inżynieria Materiałowa*, 17, 191, 1996.
27. Czyrska-Filemonowicz, A., Garbarz, B., Adrian, H., Wojtas, J. and Zielińska-Lipiec, A., Eds., *Proceedings of the IX Conference on Electron Microscopy of Solids*, Zakopane, Poland, 1996.
28. De Hoff, R.T. and Rhines, F.N., *Quantitative Microscopy*, McGraw-Hill, New York, 1968.
29. Eggleston, P., General Support Tools for Algorithm Development and Scientific Research, *Proceedings of the International Conference on Electronic Imaging*, Boston, 1991.
30. El Soudani, S.M., Profilometric analysis of fractures, *Metallography*, 247, 1978.
31. Faugeras, O., *Three-dimensional Computer Vision*, MIT Press, Cambridge, MA, 1993.

32. Favier, E. and Laget, B., A new multi-scale approach to the segmentation of multi-textured images, *Acta Stereologica*, 14, 179, 1995.
33. *Fracture*, Liebowitz, H., Ed., Academic Press, New York and London, 1968.
34. Gauthier, G., Coster, M., Chermant, L. and Chermant, J.L., Morphological segmentation of cutting tools, *Microscopy, Microanalysis, Microstructure*, 7, 339, 1996.
35. Gauthier, G., Quenec'h, J.L., Coster, M. and Chermant, J.L., Segmentation of grain boundaries in WC-Co cermets, *Acta Stereologica*, 13, 209, 1994.
36. Gentier, S., Billaux, D., Hopkins, D., Davias, F. and Riss, J., Images and modeling of the hydromechanical behaviour of a fracture, *Microscopy, Microanalysis, Microstructure*, 7, 513, 1996.
37. Gundersen, H.J.G., Bagger, P., Bendtsen, T.F., Evans, S.M., Korbo, L., Marcussen, N., Moller, A., Nielsen, K., Nyengaard, J.R., Pakkenberg, B., Sorensen, F.B., Vesterby, A. and West, J.M., The new stereological tools: Disector, fractionator, nucleator and point sampled intercepts and their use in pathological analysis and diagnosis, *Acta Pathologica, Microbiologica et Immunologica Scandinavica,* 5, 379, 1988.
38. Hanson, A. and Riseman, E., *From image measurements to object hypotheses*, COINS Technical Report 87-129, University of Massachusetts at Amherst.
39. Hazotte, A. and Lacaze, J., Quantitative analysis of the microstructure of nickel based superalloys, *Revue de Métallurgie*, 91, 217, 1994.
40. Hebdowska, M., *Application of image analysis in the evaluation of the shape of graphite precipitations in cast iron*, M.Sc. thesis, Cracow University of Technology, Cracow, 1997.
41. Heerens, J., Comec, A. and Schwalbe, K.H., Results of a round-robin on stretch zone width determination, *Fatigue Fracture Engineering Materials Structures*, 1, 19, 1988.
42. Henault, E. and Chermant, J.L., Parametrical investigation of gray tone image, *Acta Stereologica,* 11/Suppl, 665, 1992.
43. Huebner, K., Quantitative characteristics of the shape of graphite precipitates in nodular cast iron, *Przegląd Odlewnictwa*, 3, 60, 1977.
44. Huebner, K.J., Application of color metallography of microstructural images in research studies, *Proc. Q-Mat'97, International Conference on the Quantitative Description of Materials Microstructure*, Warsaw, Poland, 299, 1997.
45. *Image Analysis. Principles and practice*, Joyce-Loebl, 1985
46. Jernot, J.P., Coster, M. and Chermant, J.L., Mesure des cous dans des materiaux frittes, *Journal de Microscopie et de Spectroscopie Electroniques*, 12, 49,1987.
47. Jeulin, D., Modelisation des microstructures par des textures aleatoires, *Journal de Microscopie et de Spectroscopie Electroniques*, 12, 133, 1987.
48. Jeulin, D., Mathematical morphology and materials image analysis, *Rapport N-31/87/G, Centre de Geostatistique*, Fontainebleau, 1987.

49. Kobayashi, T. and Shockey, D.A., *Characterization of cracking behaviour using post-test fractographic analysis*, EPRI NP.-7173, Project 2614-40, Final Report, SRI International, Menlo Park, California, 1991.
50. Kohler, R., A segmentation system based on thresholding, *Computer Graphics and Image Processing,* 15, 319-338 (1981).
51. Kohler, R.R., *Integrating non-semantic knowledge into the image segmentation process*, Ph.D. thesis, Computer and Information Science Technical Report 84-04, University of Massachusetts at Amherst, 1984.
52. Kurdy, M.B. and Jeulin, D., Directional mathematical morphology operations, *Acta Stereologica*, 14, 473, 1989.
53. Kurzydłowski, K.J., Quantitative studies of recrystallization in polycrystaline materials, *Acta Stereologica*, 13, 277, 1994.
54. Kurzydłowski, K.J. and Ralph, B., *The Quantitative Description of the Microstructure of Materials*, CRC Press, Boca Raton, Florida, 1995.
55. Lantuejoul, C. and Toniolo, G., Mesure de l'allongement de particules de graphite dans des fontes, *Journal de Microscopie et de Spectroscopie Electroniques*, 12, 415, 1987.
56. Le Pennec, F. and Malewicz, D., Automatic grain size measurement in low carbon steels by image analysis, *Microscopy, Microanalysis, Microstructure*, 7, 433, 1996.
57. Maliński, M. and Chrapoński, J, 3D modelling of polycrystalline grain structures, Parts I and II, *Proc. Q-Mat'97, International Conference on the Quantitative Description of Materials Microstructure*, Warsaw, Poland, 383, 1997.
58. Maliński, M., Cwajna, J. and Chrapoński, J., Grain size distribution, *Acta Stereologica*, 10, 73, 1991.
59. Mandelbrot, B.B., Passoja, D.E. and Paullay, A.J., Fractal character of fracture surfaces of metals, *Nature*, 8, 721, 1984.
60. Marchlewski, P.A., Olszyna, A.R. and Kurzydłowski, K.J., Abnormal grain growth in sintered alumina, *Proc. Q-Mat'97, International Conference on the Quantitative Description of Materials Microstructure*, Warsaw, Poland, 397, 1997
61. Mazille, J.E., Contribution de l'analyse d'images a l'etude des processus de solidificatiom dendritique, *Journal de Microscopie et de Spectroscopie Electroniques*, 12, 57,1987.
62. Mecholsky, J.J., Passoja, D.E. and Feinberg-Ringel, K.S., Quantitative analysis of brittle fracture surfaces using fractal geometry, *Journal of the American Ceramic Society,* 1, 60, 1989.
63. Missiaen, J.M. and Chaix, J.M., Large scale point covariance analysis of homogeneity in TiB_2-Fe composites, *Acta Stereologica*, 13, 149, 1994.
64. Młynarczuk, M. and Beucher, S., Directional fibre analysis by mathematical morphology, *Proc. Q-Mat'97, International Conference on the Quantitative Description of Materials Microstructure*, Warsaw, Poland, 403, 1997.
65. Montoya, F., Michelland, S., Chermant, L., Coster, M. and Chermant, J.L., Size distribution from grey tone level images: case of granular materials, *Acta Stereologica*, 14, 401, 1989.

66. Pande, C.S., Richards, L.R. and Dempsey, B.D. and Schwoelbe, A.J., Fractal characterization of fractured surfaces, *Journal of Materials Science Letters*, 295, 1987.
67. Pierret, A. and Moran, C.J., Quantification of orientation of pore patterns in X-ray images of deformed clay, *Microscopy, Microanalysis, Microstructure*, 7, 421, 1996.
68. Pirard, E., Morphometric analysis of powders: a systematic and robust approach using mathematical morphology, *Revue de Métallurgie*, 91, 295, 1994.
69. Pratt, W., *Digital Image Processing*, John Wiley & Sons, New York, 1978.
70. Quenec'h, J.L., Chermant, J.L., Coster, M. and Jeulin, D., Example of application of probabilistic models: determination of the kinetics parameters during liquid phase sintering, *Microscopy, Microanalysis, Microstructure*, 7, 573, 1996.
71. Ralph, B. and Kurzydłowski, K.J., Microscopical quantification as an input to microscopical modeling, *Proc. Q-Mat'97, International Conference on the Quantitative Description of Materials Microstructure*, Warsaw, Poland, 141, 1997.
72. Rawicki Z. and Wojnar L., On quantitative fractographic analysis of high density concretes. *Acta Stereologica*, 2, 185, 1992.
73. Redon, C., *Morphology and mechanical behavior of amorphous cast iron fiber reinforced concrete*, Thesis of the University of Caen, France, 1997.
74. Reynolds, G., Strahman, D., Lehrer, N. and Kitchen, L., *Plausible Reasoning and the Theory of Evidence*, COINS Technical Report 86-11, University of Massachusetts at Amherst.
75. Richter, J. Cwajna, J. And Szala, J., Verification of empirical structure-properties relations through automatic image analysis of sintered carbides, *Proc. Q-Mat'97, International Conference on the Quantitative Description of Materials Microstructure*, Warsaw, Poland, 451, 1997.
76. Riseman, E., Hanson, A. and Belknap, R., *The Information Fusion Problem: Forming Token Aggregations Across Multiple Representations*, WCOINS Technical Report 87-48, University of Massachusetts at Amherst.
77. Robert, F. and Lefebvre, G., Distance mapping for image filtering, *Acta Stereologica*, 14, 209, 1995.
78. Roskosz, S. and Cwajna, J., Application of quantitative fractography methods in the investigations of drills wear mechanisms, *Proc. Q-Mat'97, International Conference on the Quantitative Description of Materials Microstructure*, Warsaw, Poland, 465, 1997.
79. Russ, J.C., *Practical Stereology*, Plenum Press, New York, 1986.
80. Russ, J.C., *The Image Processing Handbook. Second edition*, CRC Press, Boca Raton, Florida, 1995.
81. Russ, J.C. and Russ, J.C., Automatic discrimination of features in grey-scale images, *Journal of Microscopy*, 148, 263, 1987.
82. Ryś, J., *Stereologia Materiałów*, Fotobit-Design, Cracow, 323, 1995.

83. Sękowski, K. and Huebner, K., Analysis of the microscopic image with spherical precipitates on the example of nodular cast iron, *Proc. IX Seminar PAN-STOP*, Cracow, 33, 1975.
84. Serra, J., *Image Analysis and Mathematical Morphology*, Academic Press, London, 1982.
85. Serra, J., Ed., *Image Analysis and Mathematical Morphology. Volume 2: Theoretical Advances*, Academic Press, London, 1988.
86. Serra, J., Morphological image segmentation, *Acta Stereologica*, 14, 99, 1995.
87. Serra, J. and Vincent, L., *Lecture notes on morphological filtering*, Ecole de Mines de Paris, Centre de Morphologie Mathematique, Fontainebleau, 1989.
88. Sigl, L.S. and Exner, H.E., Experimental study of the mechanics of fracture in W-Co alloys, *Metallurgical Transactions A*, 18A, 1299, 1987.
89. *Standard practice for determining inclusion content of steel and other metals by automatic image analysis*, ASTM E1245-89, 1989.
90. *Standard practice for preparing and evaluating specimens for automatic inclusion assessment of steel*, ASTM E 768-80, 1985.
91. *Standard test methods for characterizing duplex grain sizes*, ASTM E 1181-87, 1987.
92. *Stereology and quantitative metallography*, ASTM STP 504, 1971.
93. Szala, J., *Principles of quantitative fractography applied in assessment of metallic materials*, Ph.D. thesis, Silesian University of Technology, Katowice 1985.
94. Szala, J. and Olszówka-Myalska, A., Application of SEM to Quantitative evaluation of multiphase materials microstructure, *Proc. Q-Mat'97, International Conference on the Quantitative Description of Materials Microstructure*, Warsaw, Poland, 523, 1997.
95. Szala, J. and Roskosz, S., Methods of the automatic twin boundaries detection, *Proc. Q-Mat'97, International Conference on the Quantitative Description of Materials Microstructure*, Warsaw, Poland, 529, 1997.
96. Tadeusiewicz, R. and Korohoda, P., *Komputerowa Analiza i Przetwarzanie Obrazów*, Wyd. Fundacji Postępu Telekomunikacji, Kraków 1997.
97. Tadeusiewicz, R., *Systemy Wizyjne Robotów Przemysłowych*, WNT, Warszawa 1992.
98. Talbot, H., The pre-processing in mathematical morphology, *Revue de Métallurgie*, 91, 211, 1994.
99. Talbot, H., Jeulin, D. and Hanton, D., Image analysis of insulation mineral fibres, *Microscopy, Microanalysis, Microstructure*, 7, 361, 1996.
100. Underwood, E.E., *Quantitative Stereology*, Addison Wesley, Reading, MA, 1970.
101. Underwood, E.E. and Banerji, K., Quantitative fractography, in: *Metals Handbook*, Ninth Edition, vol. 12, Metals Park, Ohio, 193, 1987.

102. Vincent, L. and Jeulin, D., Minimal paths and crack propagation simulations, *Acta Stereologica*, 14, 487, 1989.
103. Watkins, C.D., Sadun, A. and Marenka, S., *Modern Image Processing: Warping, Morphing and Classical Techniques*. Academic Press, 1993. Polish edition: WNT, Warsaw, 1995.
104. Weibel, E.R., *Stereological Methods*, Academic Press, London, 1980 and 1989.
105. Wendrock, H. and Huebel, R., Characterization of microstructural anisotropy of steels by means of mathematical morphology, *Acta Stereologica*, 13, 143, 1994.
106. Wojnar, L., *Effect of graphite on the fracture toughness of a nodular cast iron*, Ph.D. thesis, Cracow University of Technology, Cracow 1985.
107. Wojnar L., *Fraktografia Ilościowa. Podstawy i Komputerowe Wspomaganie Badań*, Cracow University of Technology, 1990.
108. Wojnar, L., 10 years of quantitative fractography development (1983-1993) , *Inżynieria Materiałowa*, 14, 89, 1993.
109. Wojnar, L. Ed., *Proc. STERMAT'94. IV International Conference: Stereology and Image Analysis in Materials Science*, Beskidy Mountains, Poland, 1994.
110. Wojnar, L., *Quantitative analysis of the fracture surfaces*, Research grant of the State Committee for Scientific Research No 3-0011-91-01, 1992.
111. Wojnar, L., Modern approach to microstructure image analysis, *Proc. Q-Mat'97, International Conference on the Quantitative Description of Materials Microstructure*, Warsaw, Poland, 103, 1997.
112. Wojnar, L. and Kumosa, M., Quantitative analysis of overlapped fracture surfaces, *Engineering Fracture Mechanics*, 4, 597, 1990.
113. Wojnar, L. and Latała, Z., On the validity of specimen preparation in automatic analysis of the microstructure of cast iron, *Proc. First Conference: Materials Science - Foundry - Quality*, Cracow, 165, 1997.
114. Wojnar, L. and Majorek, M., *Komputerowa Analiza Obrazu*, Fotobit-Design, Cracow, 1994.
115. Wojnar, L. and Szala, J., Fraktografia ilościowa, in: Ryś, J., *Stereologia Materiałów*, Fotobit-Design, Cracow, 323, 1995.
116. Young Tzay, Y., and Fu King-Sun, *Handbook of Pattern Recognition and Image Processing*, Academic Press, San Diego, 1986.
117. Zielińska-Lipiec, A., Adrian, H., Jennis, P. and Czyrska-Filemonowicz, A., Quantitative comparison of the microstructure of high chromium steels for advanced power stations, *Proc. Q-Mat'97, International Conference on the Quantitative Description of Materials Microstructure*, Warsaw, Poland, 565, 1997.
118. Zou, X., *Electron Crystallography of Inorganic Structures*, Stockholm University, Stockholm, 1995.

Subject index

C

D

G

P

Q

R

S

T

U

V

W

X

Z